MICROSCOPY HANDBOOKS 24

The Preparation of Thin Sections of Rocks, Minerals, and Ceramics

D. W. Humphries
Formerly Senior Lecturer, Department of Geology, University of Sheffield

Oxford University Press · Royal Microscopical Society · 1992

Oxford University Press, Walton Street, Oxford OX2 6DP
Oxford New York Toronto
Delhi Bombay Calcutta Madras Karachi
Petaling Jaya Singapore Hong Kong Tokyo
Nairobi Dar es Salaam Cape Town
Melbourne Auckland
and associated companies in
Berlin Ibadan

Royal Microscopical Society
37/38 St Clements
Oxford OX4 1AJ

Oxford is a trade mark of Oxford University Press

Published in the United States
by Oxford University Press, New York

A catalogue record of this book is available from the British Library

Library of Congress Cataloging in Publication Data
Humphries, D. W.
The preparation of thin sections of rocks, minerals, and ceramics
/D. W. Humphries.
p. cm.—(Microscopy handbooks; 24)
Includes bibliographical references and index.
1. Thin sections (Geology)—Laboratory manuals. 2. Microscope and microscopy—Technique—Laboratory manuals. I. Title. II. Series.
QE434.H86 1992 552'.0028—dc20 91-33027
ISBN 0–19–856431–7

Typeset by Cotswold Typesetting Ltd, Cheltenham
Printed in Great Britain by St. Edmundsbury Press,
Bury St. Edmunds, Suffolk

Preface

Comparatively little can be learned of the structure of rocks and minerals from the examination of fractured surfaces by reflected light. Flat polished surfaces show very much more, but nearly all the important facts can only be observed by examining thin sections by transmitted light . . . The details of the method of preparing these must necessarily vary according to the mechanical means at the disposal of each person, and much time may be saved by the use of machinery. I shall therefore give such a general account as may be used by anyone who has not machinery at command, premising that it will be easy to modify it in detail, according to the facilities which each may possess for employing more expeditious methods.

Henry Clifton Sorby, FRS (1868)
President, Royal Microscopical Society 1875–7

Had Sorby been alive just a few years ago he would have doubtless been gratified to see the Sunday supplements full of colour pictures of rocks brought back from the Moon. While many readers were doubtless attracted by the brilliance of the colours and the abstract patterns, few realized that these were, in fact, 'thin sections' 30 μm in thickness photographed by polarized light and made by methods which he had pioneered as long ago as 1849.

That rocks, minerals, and ceramics can be made thin enough to be examined by transmitted light under the microscope comes as a surprise to many people, including experienced microscopists. The art of making such sections is well within the grasp of anyone with a modicum of manual dexterity, a certain amount of 'elbow grease', and a measure of patience. Above all, they can be made without an extravagant outlay on expensive equipment. Nevertheless, Sorby's forecast that the use of machinery could save time and effort has proved correct and such methods are briefly discussed.

This book is essentially a practical guide to the making of thin sections of rocks, minerals, and ceramics, aimed on the one hand at those who already have some experience of studying them under the polarizing microscope, and who want to know how such sections are made and how they can make their own thin sections. On the other hand, it is hoped that those who know nothing of such things and are intrigued by the possibility that rocks and ceramics can be examined under the polarizing microscope will be encouraged to learn a new skill and to apply the methods to their own particular interests.

Tealby, Lincs D. W. H.
May 1991

Contents

Safety

Attention to safety aspects is an integral part of all laboratory procedures and both the Health and Safety at Work Act and the COSHH regulations impose legal requirements on those persons planning or carrying out such procedures.

In this and other Handbooks every effort has been made to ensure that the recipes, formulae, and practical procedures are accurate and safe. However, it remains the responsibility of the reader to ensure that the procedures which are followed are carried out in a safe manner and that all necessary COSHH requirements have been looked up and implemented. Any specific safety instructions relating to items of laboratory equipment must also be followed.

1 Introduction

To most microscopists, the (light) microscope is simply an instrument capable of making visible minute detail that is not seen with the naked eye. As such it has been used to examine complete small objects, mostly from the plant and animal kingdoms, crushed fragments of larger objects, and slices cut with a knife, a razor, or a microtome.

When the microscope was first applied to the study of rocks and minerals, it was possible only to examine crushed fragments, and what was seen was not very exciting—merely glassy fragments varying in colour but with little or no obvious structure. Even now, few users of the microscope realize that the microscope, with the addition of polarizing equipment and a rotating stage, is one of the most powerful tools available for the examination of these materials.

For those unfamiliar with the polarizing microscope, it is recommended that they turn to Appendix 1 for a brief account of a simple method of converting a standard 'biological' microscope to use polarized light and the preparation of a microscope slide which will illustrate the power of polarized light in discriminating between some familiar materials.

For the geologist, and for the ceramicist, the polarizing microscope is the primary instrument used in the detailed study of rocks and ceramics. With it, he or she is able to identify the components of the rock or ceramic, the size of the various particles, the relationship between them, and their distribution within the material. But, without the thin section, little of this would be possible. While it is assumed that the reader of this book is already familiar with these things and needs no further guidance, there may be some who may be attracted to the notion of making a thin section and would wish to know what they are likely to see when it is put under the polarizing microscope. For them, a few books are listed on p. 79, but the shelves of a good library or a scientific bookshop may prove to be the best starting point.

Though this handbook is intended to be a practical guide to the preparation of thin sections of rocks, minerals, and ceramics, a brief note on the history of slide making is warranted by the profound effect that the study of thin sections has had on the understanding of the nature and origin of rocks and minerals and consequently on the understanding of the history of the Earth, the Moon, and meteorites.

Two problems hindered the application of the microscope in this field of investigation. The first was the apparent impossibility of making thin slices of rocks comparable to the thin sections used by the biologist, and the second was the failure to realize the significance of the observation (made on crushed fragments) that mineral particles were nearly always *anisotropic**—a feature which could only be seen in polarized light.

In the latter part of the eighteenth century, thin plates of agates and fossil wood were made by William Nicol, who claimed to be the originator of the technique. A similar method was applied to the sectioning of teeth by Mr Nasmyth, a London dentist (Sorby 1882).

The invention of the polarizing microscope is usually attributed to David Brewster, around 1815, but in his 'Treatise' (which is dedicated to Fox Talbot) he appears to acknowledge Fox Talbot as the first to set up a compound microscope for the express purpose of examining structures by polarized light (Brewster 1837). The application of Nicol's prism to the microscope certainly appears to be due to Fox Talbot (Fox Talbot 1834*a,b*).

The use of this new instrument seems to have been applied to the examination of every slide in the cabinet of the microscopist of the period. Among the objects examined were undoubtedly the sections made by Nicol and Nasmyth as well as thin plates of minerals such as selenite and tourmaline, but without an understanding of the reasons for the often brilliant colours produced, the habit quickly died out and, indeed, the provision of polarizing equipment by the microscope makers virtually ceased.

The vital step that led to the study of rocks by means of the polarizing microscope was made by Henry Clifton Sorby (1826–1908) of Sheffield following a chance meeting on a train journey from Scarborough to York with a Mr Williamson, who later became a professor at Owen's College in Manchester. Williamson, who was then a practising surgeon, showed Sorby how to make thin sections of fossil wood, teeth, scales, and bones. Sorby, already interested in geology and destined to become President of the Royal Microscopical Society and one of the great scientists of the nineteenth century, immediately recognized the possibility of applying these new-found skills to rocks and, in 1849, produced the first (Fig. 1) of around 1000 thin sections of rocks† of a quality comparable to any that have been produced since.

He was already well versed in the theory of polarized light and in 1851 published the first account of the mineral composition of a rock, differentiating between agate (essentially quartz) and calcareous spar (calcite), which are both colourless in ordinary light, by their different appearance in polarized light (Sorby 1851). In so doing, he laid the foundations of microscopical petrology and by applying his methods to iron and steel (for examination by reflected light) laid the foundations of metallography. His researches, spanning 60 years, were published in more than 200 papers.

* A glossary of some of the terms used is given on p. 76.
† Now preserved in the Geology Department, University of Sheffield.

Fig. 1. Thin section of a rock made by H. C. Sorby in 1849 inscribed *Wenlock lim. nr. Ledbury no.1. H.C.S. 1849.*

Sorby gave a brief outline of his technique of thin section making in 1868 and a more detailed account in 1882 (Sorby 1868, 1882). A summary of his methods and an outline of the methods used today are given below.

1.1. Outline of the method

1. The initial step is the collecting of the rock or mineral specimen and breaking or sawing from this a chip about 25 mm (1″) square and as thin as possible. Sorby preferred to collect his specimens *in situ* to ensure that he was familiar with the stratification of the rock. If he was forced to use museum specimens, he would cut them with a 'slitting-saw'. It is not clear, however, whether this was a toothless hand saw of sheet iron used with an abrasive (probably emery), or a treadle-powered circular saw with crushed bort (industrial diamond) mixed with oil applied to the edge of the saw. Nowadays, slices are generally cut with a diamond-edged circular saw, but if this is not available, chips flaked off with a hammer are still a perfectly satisfactory starting point.

2. One side of the slice was then ground perfectly flat on a smooth slab of Coal Measure sandstone using emery powder and water as an abrasive. The sandstone slab tended to wear hollow with grinding slides on it and Sorby replaced it with a plate of zinc that had been hammered flat. Later, he used Congleton stone and Water of Ayr stone, with water as a lubricant, but without any added abrasive. He was particularly concerned that the

slice should be perfectly flat, that no grain of harder material stood proud of the surface and that no grain should be pulled out. Nowadays, this grinding is done on a horizontal rotating iron lap charged with carborundum powder and water, or by hand on a steel plate. The slice is then smoothed on a glass plate using finer abrasive.

3. The prepared slice is fixed to a piece of glass. Sorby used small pieces of glass 40 mm ($1\frac{5}{8}''$) square, the size being determined by the size of a polariscope which he had constructed. Having had boxes made to hold these glasses it was natural for him to make all his slides on this size of glass. Standard 76 × 26 mm (3″ × 1″) microscope slides are now usually used, but 50 mm (2″) squares of plate glass are sometimes preferred.

The adhesive used was Canada balsam—the thick viscous variety, not the solution in xylene used by biologists. Sorby gave detailed instructions for its use and these will be described later. Various alternatives, including other thermoplastic resins, epoxy resins, and photocements are now used.

4. The next step was to reduce the thickness of the portion of rock thus fastened down. Sorby placed them in the hand of a 'very intelligent glass cutter, and he used to grind them down till he left them about the thickness of a good stout card—not thinner for fear of portions being torn up and damaged'. To avoid scratching the glass, Sorby fixed small pieces of thin sheet zinc at the corners of the slide with Canada balsam. The slice was finished by grinding on the Water of Ayr stone to the desired thickness. Rocks with a moderate grain size and containing quartz were ground to about 30 μm, limestones were left rather thicker, while fine-grained rocks were made much thinner.

Nowadays, the mounted slice is ground down on rotating laps and finished on glass plates using progressively finer grades of carborundum until the section is 30 μm thick. The thickness is determined under the polarizing microscope using the interference colour of quartz, if it is present, or some other readily identifiable mineral. A preliminary (literally) rule of thumb for estimating thickness is to hold the section towards the light: if the silhouette of the fingers or thumb is visible, then the section is approximately 30 μm thick.

As Sorby predicted, more expeditious methods can be used, particularly in large workshops. Very large diamond-edged wheels can be used for cutting specimens into suitable slabs, with small saws running at very high speeds to cut the individual slices. Horizontal milling machines, with power feeds and a diamond-edged wheel, have been adapted to grind many slides simultaneously. Precision lapping and polishing machines, originally developed for use in the semiconductor industry, have been successfully applied to the preparation of thin sections.

5. If thermoplastic resins such as Canada balsam are used to mount the

slice, it is possible to transfer the finished section to a clean unscratched slide, but this is not possible with the epoxy rcsins and photocements. The finished section is usually covered with a glass cover slip and cleaned and labelled.

Modern methods of slide making are essentially the same as those devised by Sorby almost a century and a half ago. The main differences are in the types of abrasive used. He used emery, whereas carborundum and sometimes diamond are used nowadays, and he did believe in getting others, including his mother, to do much of the coarse grinding!

When one first embarks upon making thin sections there are inevitable mishaps: sections inexplicably disappear at the last moment, or they are wedge shaped, or the slice comes off the glass. Despite this, the beginner should take heart. A complete novice (*not* the author) has been known to produce a perfect thin section at only the second attempt.

2 Collecting and preparing specimens

There are a variety of reasons for collecting and sectioning a specimen. It may be to determine its mineralogy and hence its precise identity, or it may be to establish its grain size or inter-grain relations, or the conditions of its formation (manufacture, in the case of ceramics), and its subsequent history. Hence it is preferable to collect material at source whenever possible, although some difficult-to-obtain materials can be purchased from dealers in rocks, minerals, and fossils.

2.1. Collecting

Whatever the source of the specimen to be sectioned, the circumstances should be fully recorded. For rocks and minerals, this means a detailed record of the location of the exposure, its nature and extent, the uniformity or otherwise of the beds exposed, and the precise position of the specimen collected. For ceramics, it requires as much information as possible of the raw materials and the conditions of firing.

In the field, specimens should be given temporary labels, e.g. numbered metal tags, paper labels, or numbered tickets, and placed in canvas or plastic bags. All specimens must be labelled, usually by means of a number painted on them (corresponding to a number in the note book) as soon as possible after collection. They should then be wrapped in paper or placed in cardboard trays and stored in drawers. However good one's memory, a pencilled note on the edge of the paper is definitely not the best means of long term identification.

Unless a power saw is available for slicing the hand specimen, it is advisable to flake off pieces of rock at the point of collecting, using the sharp edge of the square end of a geological hammer. These pieces should be about 25 mm (1″) square and as thin as possible (about 6 mm ($\frac{1}{4}$″) thick). For bedded rocks, it is best to break the flakes across the stratification. For massive rocks, this is less important, but it is still desirable to know the orientation of the flake in relation to the main body of the rock, if that is possible. For rocks which are strongly foliated, it will usually be necessary to break off thicker flakes across the foliation as well as to break flakes along the foliation. For ceramics, the size and nature of the items will obviously control the possibility of flaking off sections as described above. More usually, sections will have to be sawn from the original sample.

2.2. On the use of hammers

It is imperative to use a hammer made specifically for breaking rocks or rock-like materials. The first choice is a geological hammer with a square head and a chisel-edged pane, about 1 lb (500 g) in weight. An alternative is a bricklayers' or slaters' hammer. The square head should always be used for breaking rocks (never the chisel end) and the blow should be struck away from the user or downwards and not towards any other person. Flakes of rock have a habit of flying considerable distances.

For added protection, goggles should be worn. It should be noted, however, that soft plastic goggles or eyeshields should never be worn over spectacles, since there is a danger that a flying splinter of rock could depress the plastic and fracture the glass behind it. Only goggles with unbreakable, rigid lenses should be used.

In no circumstances should an engineers' or carpenters' hammer (or even the old-fashioned coal hammer) be used since these are liable to shed very fine splinters of steel when struck on rock.

Finally, it is desirable to keep the use of the hammer to a minimum in the countryside. In recent years, enormous damage has been done to important exposures by uncontrolled hammering by large parties with the consequence that access to land and the removal of rocks and fossils has been prohibited by some landowners. Indiscriminate hammering must be avoided at all times.

2.3. Rock-breaking machines

Large pieces of rock and especially cores of rock from borings are most easily broken in breaking machines. Unfortunately, such machines appear to be no longer available on the market, but with a little ingenuity suitable devices can be fabricated fairly easily.

The simplest rock breaker is a large engineers' vice, in which the flat jaws have been replaced by chisel-shaped jaws. An older form of screw-type breaker is shown in Fig. 2. A hydraulic car jack can be used as the ram for the moving jaw (Fig. 3), but it should be noted that car jacks cannot be inverted since the valves operate under gravity.

The jaws in all these machines have an included angle of about 105° and are hardened to the same degree as a cold chisel. It must be remembered that considerable forces can be exerted in both the screw and the hydraulic type of machine and that adequate metal must be allowed in the various components. In all types, protection from flying rock fragments must be provided.

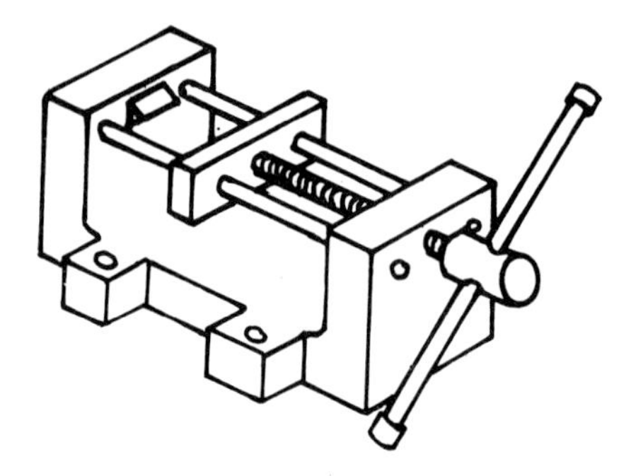

Fig. 2. Screw-type rock-breaking machine.

Fig. 3. A 'home-made' hydraulic rock-breaking machine, using a bottle jack and a heavy steel frame. Note the hinged, wire-mesh grill to protect the operator.

2.4. Rock saws

Four types of rock saw are, or have been, used for cutting thin slices prior to grinding. The oldest type had a soft iron or copper disc, usually about 225 mm (9″) in diameter rotating on either a vertical or horizontal spindle and armed with finely crushed bort (industrial diamond). The diamond was made into a paste with oil and applied to the edge of the wheel which was set to run slowly. A piece of flint pressed against the edge of the wheel forced the diamond dust into the disc. The rock to be cut was fixed in a clamp and held firmly against the edge of the wheel, water being used as a lubricant. The earliest type of machine was treadle driven, with an electric motor ultimately replacing muscle power. Carborundum powder can be made into a paste and fed on to the edge of the wheel, but the cutting is slow and not as 'sweet' as with diamond.

Modern rock-cutting machines (Fig. 4) use either a steel disc with a copper edge which is impregnated with diamond or a thin disc made of abrasive grit. A fourth type of machine is similar to a bandsaw but uses a diamond-impregnated wire. These saws are usually lubricated and cooled with water.

Makers of cutting discs generally recommend cutting speeds which are rather higher than is desirable: lower speeds will avoid strain within the specimen which could lead to fracturing of grains.

Fig. 4. A modern rock-cutting machine using a diamond-impregnated blade. The specimen can be clamped in the vice attached to the rail on the right-hand side of the machine; mounted slices can be held in the vacuum chuck on the left of the cutting wheel. A Perspex hood protects the operator and reduces noise. (Photograph by courtesy of Logitech Ltd.)

A simple rock-cutting machine, based on a standard bench grinder and using a copper or iron disc, a diamond edged wheel, or an abrasive disc is not difficult for the amateur engineer to construct. Great care should be exercised in providing adequate protection for the operator, especially if abrasive wheels are used. The specimen may be hand-held on a table designed to travel parallel to the plane of the wheel or it may be held in a pivoted clamp.

3 Initial grinding

By now the slide maker should have a sawn slice or a chip of the specimen to be sectioned. Some materials are extremely fragile or friable and it may have proved impossible to obtain either the slice or the chip. Such materials will need special treatment, such as consolidation or impregnation. This will be dealt with later.

The next step in the preparation of the thin section is to smooth the surface of a sawn slice or to grind a flat surface on a chip. This can be done either by hand on grinding plates or by means of a power grinder using carborundum powder and water as the abrasive.

3.1. Abrasives

There are a number of abrasive materials which are suitable for grinding rocks including diamond, emery and the artificial materials, carborundum and boron carbide. Diamond is used in some cutting and grinding wheels. It is also used for fine grinding and polishing. Emery, which is essentially corundum (Al_2O_3) has been used extensively in the past as an abrasive and is still used to some extent for fine grinding and as a polishing powder. Most of the material used is, however, a synthetic product and not natural emery. The most commonly used abrasive is carborundum (SiC), made by fusing sand and coke in an electric-arc furnace. The fused product is then crushed and graded. The virtues of carborundum are that it is hard (9 on Moh's scale), it resists attrition, and it breaks with a conchoidal fracture so that grains always have a sharp cutting edge. Boron carbide is almost as hard as diamond and is used particularly in the grinding of very hard ceramics.

Three grades of carborundum are normally used in slide making by hand methods: coarse, 120 mesh (100 μm); intermediate, 220 mesh (60 μm); and fine, 3F (12 μm). The development of chucks for holding slices and slides has led to the use of 600 mesh (20 μm) carborundum on rotating grinding laps. Although this is a very slow process, it minimizes damage to the specimen resulting from the use of coarse powders. Nevertheless, satisfactory slides have been made for many years, using the grades recommended above.

It is essential to keep the various grades of carborundum completely separated and coarser grades should never be allowed to contaminate finer grades. In use, the standard containers (500 g, 1 lb) are rather too large to allow easy handling and small, cheap pepper pots with perforated screw-on caps are a convenient means of dispensing the various grades.

3.2. Grinding plates

The plates for hand grinding are either steel or glass, 300–375 mm (12–15″) square. Steel plates are usually about 10 mm (½″) thick and should have a machine-ground surface, so that they are perfectly flat. Only one steel plate is generally used (for coarse grinding) and that is not always regarded as essential, although it is more resistant to grinding with coarse carborundum than glass.

Glass plates are usually about 6 mm (¼″) thick and are readily obtainable from glaziers and many DIY or hardware stores. The cut edges are likely to be very sharp and some suppliers will grind them, but it is a useful exercise to smooth them oneself and thereby get a 'feel' for using carborundum. The simplest method is to pat a piece of wetted wood about 75 × 50 × 25 mm (3″ × 2″ × 1″) on to a small heap of 120 grade carborundum and with the glass plate just overhanging the edge of the workbench, smooth off the edges by firm strokes of the wood along the length of the edges. Alternatively, the edges can be smoothed off with a small, flat carborundum stone (of the type used for sharpening woodworking tools) wetted with water. It is unnecessary to produce a perfectly smooth finish. All that is required is to remove the edges and corners on which the fingers could be cut when the plates are in use.

Three plates in all are required, one for each grade of carborundum. With care in use, grinding plates should remain flat for a long time. When too badly worn, they can be turned over and when the second side has worn, it is probably easier to replace them rather than going to the trouble of grinding them flat.

Although thick glass plates are preferable, ordinary window glass can be used, but glass sold as 'horticultural glass' should be avoided since the surface is often slightly rippled.

3.3. The workbench

The grinding plates should be used on a substantial bench or table so that there is no possibility of the plates flexing as the specimen is ground. It is sometimes recommended that each plate is placed in a separate tray. Tin tea-trays will suffice provided they are perfectly flat. Rather cheaper and just as effective are several sheets of newspaper spread over the bench, to absorb the water that runs off the plates and to collect the slurry produced by grinding. The paper can be replaced as frequently as required. The plates should be in line and about 300 mm (1′) or so apart to minimize the chances of carborundum powder being carried from one plate to the next.

It is highly desirable to have ready access to running water for washing slides as they are being ground and to clean up plates at the end of the grinding session. However, it is advisable to have a large sump beneath the sink to trap the solid matter washed off to avoid blocking waste pipes. If running water is not available close to the grinding bench, small basins of water adjacent to each plate can be used to rinse off the slide as grinding progresses, but thorough washing before moving from one plate to another is essential. Hands should also be washed thoroughly to avoid the transfer of carborundum from one plate to the next.

The bench should be long enough to have space for a polarizing microscope and lamp close to the final grinding plate in order to check the interference colours (and hence the thickness) of the minerals of the slice being made. Great care must, of course, be taken to avoid getting any of the carborundum on to the microscope and it may be preferable to have the microscope on a separate bench, but this should be close enough to the grinding area to allow frequent checking in the final stages of grinding.

Working space must also be found in the workshop for slide mounting and covering.

3.4. Grinding machines

If only a small number of slides is to be made, grinding machines are not essential, though there is no doubt that they can speed up the process of slide making, as well as save a great deal of manual labour.

The simplest type of grinding machine is essentially a circular iron plate about 300 mm (12″) in diameter and 25 mm (1″) thick, mounted at the top of a vertical spindle and rotated at about 50–60 r.p.m. This spindle is mounted in the floor of an open-topped box which acts as a trap for the carborundum and sludge which is thrown off the disc as grinding proceeds. The power source is an electric motor (of about $\frac{1}{4}$ h.p. or 186.5 W) with speed reduction through a gearbox or, more often, a system of pulleys. For safety, the motor switch is placed at floor level and is foot operated. A machine of this type (Fig. 5) can readily be constructed in a moderately well equipped workshop. Twin-disc machines are not common, but do have the advantage that both coarse and intermediate grinding can be carried out on them.

It is not practicable to use one disc for more than one grade of carborundum, since it is a highly laborious task to fully clean the machine before changing the grade. Usually, the single disc will be used for coarse grinding if flakes are used as the starting material, or for intermediate grinding if sawn slices are available.

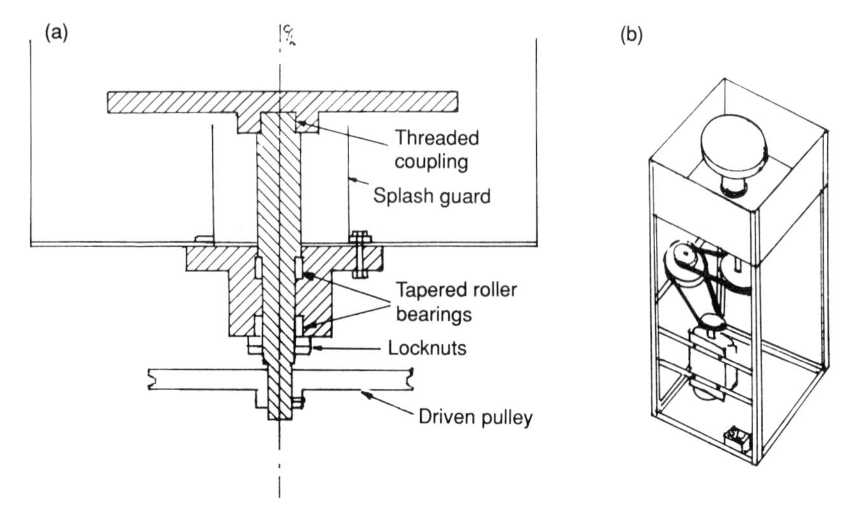

Fig. 5. (a) Schematic diagram of the grinding plate and bearings of a simple grinding machine; (b) sketch of complete machine—this is free standing and about 1 m (39″) high.

Modern, commercially made grinding machines are equipped with chucks and carriers for grinding both rock chips and sawn slices, as well as feeders for lubricants and abrasives. These machines can grind or polish up to four samples simultaneously and can be left unattended for long periods (Fig. 6).

Very large machines, capable of grinding 10 to 15 slices simultaneously, have been constructed. They were generally converted horizontal or vertical milling machines with a wide carborundum- or diamond-edged wheel or cup mounted on the spindle. The machine table was traversed automatically, and movement upwards as grinding progressed was by manual or automatic feed.

A surface grinder was adapted (Jones and Hawes 1964) to grind very large thin sections of concrete up to 120 mm (5″) square. The author recalls seeing, many years ago, thin sections of concrete 300 mm (12″) square which were probably produced on this type of machine. In this connection, it is interesting to note that thin sections of 'coal balls' (nodules containing plant remains from coal seams) of similar size were at one time commonly made by purely manual methods.

If the amateur slide maker is tempted to indulge in this degree of mechanization, it is likely that he or she will have sufficient mechanical ability to appreciate that there are a number of problems to be overcome and hopefully, will have the ingenuity to contrive solutions to them. The professional (University or commercial) slide maker, once the principal converter of machine tools, is now likely to use the machines developed for the electronics industry for the manufacture of wafers for semiconductors, substrates for electro-optic and acousto-optic materials, and mirrors,

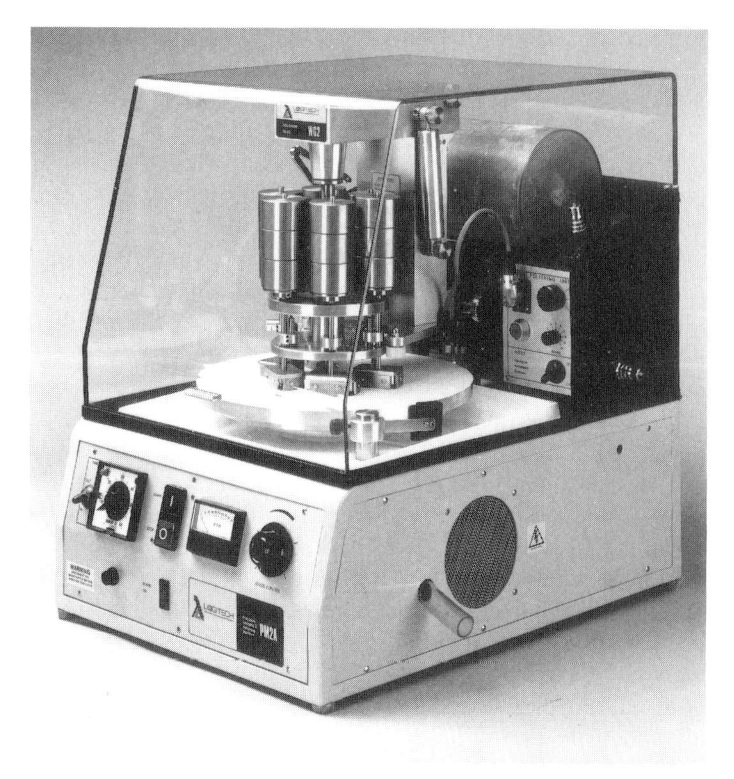

Fig. 6. A grinding and polishing machine with multiple chuck for grinding five specimens simultaneously. (Photograph by courtesy of Logitech Ltd.)

windows, and prisms for lasers. The lapping machines are essentially similar to the simple grinding lap described earlier, but have a wide range of lapping discs and sophisticated chucks for holding the specimens. By using automated methods of dispensing abrasives, very rapid processing can be achieved (Fig. 6).

3.5. Hand grinding

If a slice has been sawn, only smoothing with 3F grade carborundum will be required, but if a chip is to be ground, more vigorous treatment will be needed. The first step is to sprinkle a little of the coarse (120 grade) carborundum on the plate and wet it with a little water from a plastic wash bottle, a spray bottle such as those used for dispensing a number of domestic products, or simply dripped from a bowl with the fingers. The amount of water to be added comes with experience, but only a little is required—enough to dampen the powder and not enough to flood it. The flatter side of the chip is then placed on the carborundum and, with a circular motion applied with the fingertips, the grinding can begin. Grinding should be slow and deliberate and always with a circular movement. As rock flour accumulates, further small amounts of water can be added to keep the abrasive cutting. If too much water is added the grains of carborundum tend to roll out from under the chip and there is no appreciable cutting action. If too little water is added, the chip will tend to stick to the plate. The aim is to produce a thin slurry of carborundum, rock flour, and water in which the chip will slide and at the same time be ground down. That grinding is occurring can be sensed through the fingertips and heard as a 'grinding noise'. Pressure on the chip needs to be steady, but not excessive and the circular grinding motion must be spread progressively over the plate, right up to the corners, so that wear is uniform. The use of both hands is less tiring than using first one hand and then the other and furthermore gives much better control over the movement of the chip. Grinding should continue until a flat face is produced over as large an area of the specimen as possible. During this period of grinding, which can be quite time-consuming, more water and more abrasive can be added periodically. Any temptation to 'scrub' the chip up and down the plate must be avoided since this is liable to cause uneven wear. If grinding has to be stopped because of fatigue or other reason (and with some rocks, e.g. flint and chert, grinding may take many hours) the slice should be removed from the plate, washed thoroughly and laid aside, ground surface uppermost.

When all traces of the original fractured surface have been ground away, and the flat surface is free from visible pits and hollows, the specimen (and the hands) should be scrubbed clean under running water, using a nail brush, to remove all traces of carborundum and rock flour. Grinding can now continue exactly as before with the intermediate grade of carborundum (220 grade). After a few minutes grinding, the surface should be washed, wiped dry on a clean cloth, and examined carefully. Most specimens will show a pattern of scratches and sometimes cracks which were produced in the first stage of grinding. These must be removed by continuing the grinding on the intermediate grit. Care must still be taken to apply uniform pressure to the slice and to maintain the regular circular motion over the whole area of the plate. Straight-line grinding, with pressure alternately on the leading edge of the slice as the direction of grinding changes, is very likely to lead to the development of a curved surface. The surface must at all times be kept flat. It should be noted that the 220 grade of grit cuts quite rapidly and particular care should be taken to keep control of the grinding process by guiding the specimen and not forcing it.

When the slice is free from all the scratches produced by the coarse grit, it is again thoroughly cleaned and transferred to the third plate for grinding with the fine carborundum (grade 3F). The plate is wetted and carborundum powder is sprinkled on to it. The slice is ground, still with a circular motion, until a very fine matt finish is produced. It is given its final scrubbing and washing, wiped dry on a clean cloth, and set aside to dry. Wiping on paper towels is not recommended because fine fibres are easily left adhering to the specimen, and these are all too readily visible in the finished thin section when viewed between crossed polars.

If, at any stage, scratches begin to reappear on the ground surface, the slice and the plate should be washed clean and grinding re-commenced with fresh carborundum. It may even be necessary to go back to a coarser grade, if serious contamination of a plate with coarse particles has occurred.

With experience, it will be found that some materials can be ground satisfactorily using only the intermediate and fine grades of carborundum. Such rocks are generally fine grained and include basalts and some limestones.

It is essential that this initial grinding of the specimen produces a truly flat surface. If it does not, no amount of careful final grinding will produce a satisfactory thin section. If the initial surface is not flat it is likely to be convex and the final section will be thinner at the edges than at the centre and the section will rapidly decrease in size as grinding proceeds. Finished sections which tend to be circular are a sure sign that the initial grinding did not produce a truly flat surface (see Fig. 8).

3.6. Machine grinding

Grinding on a simple power-driven disc differs slightly from hand grinding. The disc should be set running and wetted with water. The carborundum grit (whether coarse or intermediate) is then placed near the centre of the disc in a small pile and the wetted specimen applied to the edge of the pile to pick up the grit and gradually spread it over the whole of the disc. Instead of the circular grinding motion, as in hand grinding, the specimen is moved from side to side across the diameter of the disc. Sufficient water to produce a slurry is necessary to prevent the chip being pulled out of the hands and thrown into the surrounding box. It should be remembered that the specimen is being alternately pulled and pushed into or out of the hand as the wheel rotates.

Whether one or two power-driven discs are used, the final smoothing on the fine carborundum plate is normally done by hand, as before.

Power grinding probably uses more carborundum than hand work and there may be a temptation to recover carborundum from the sludge that accumulates in the surrounding box. This is possible with great care in washing the sludge but the time and labour involved and the danger of not recovering perfectly clean grit, free from coarser particles (often of rock broken from the specimens being ground), makes the practice inadvisable. It is best to allow the sludge to dry out and dispose of it in the waste bin.

With semi-automatic grinding machines in which the slices or chips are held in a chuck, it is usual to omit the coarse and intermediate grinding powders and use only the 3F or 600 grade, which is dispensed automatically to the disc in the form of a suspension. Grinding is continued for at least 30 minutes to produce a truly flat surface in which no shattering of grains has occurred. Grinding times can be reduced considerably by the use of diamond slurries. The high cost of diamond abrasive is said to be offset by the increased productivity attainable.

4 Mountants and mounting

The next step is the mounting of the prepared slice on to a suitable support for the final stage of grinding.

The usual support for the slice is a standard 76 × 26 mm (3″ × 1″) microscope slide, though some workers prefer to use a 50 mm (2″) square piece of plate glass 6 mm ($\frac{1}{4}$″) thick. The sections made by Sorby were mounted on thin glass about 40 mm (1$\frac{5}{8}$″) square (see p. 4). Continental slide makers have also used squares of glass of about this size. Since the modern microscope and the associated mechanical stages are adapted to 76 × 26 mm (3″ × 1″) slides, it is necessary to complete the slide-making process by transferring the section from the 'odd' sizes of glass to the standard slide. This is possible, however, only if the mounting medium can be liquefied when the grinding is finished. With modern cements, this is not possible and consequently, the slide on which the slice is initially mounted will be the one on which the section is finished, and used. An exception to this arises in the making of polished thin sections (Chapter 12).

4.1. Mounting media and their use

The principal requirements of a mounting medium are strong adhesion to glass and to the rock-forming minerals, high strength in shear, absence of colour, long-term stability, and high refractive index. The most commonly used materials include thermoplastic resins and epoxy resins. The cyanoacrylate cements (superglues) will certainly hold rock slices on to glass but their adhesion to glass is probably lower than that of the more traditional cements. Photocements have not been used for slide making, but do appear to have considerable potential since they adhere strongly to glass and, once cured in ultraviolet light, leave no sticky residues.

4.1.1. Thermoplastic resins

These include the natural resin, Canada balsam, and a synthetic resin, known by its commercial name, Lakeside 70C.

An essential piece of equipment required when using thermoplastic resins is a **hotplate**. This may be a commercial electric hotplate with thermostatic control, a tripod stand with a metal plate on it heated with a Bunsen burner or spirit lamp, or simply a metal plate placed on the hob of a gas or electric cooker. The metal plate should be either copper, brass, or aluminium, about 6 mm ($\frac{1}{4}$″) thick and 150–225 mm (6–9″) square. A simple hotplate can be improvised by inverting an electric laundry iron fitted with a thermostatic control. The maximum temperature required for heating the thermoplastic cements is no more than about 140 °C. The temperature can be measured with a mercury thermometer inserted into a hole in a short length of brass rod about 25 mm (1″) long by 25 mm (1″) diameter standing on the hotplate. A thermocouple cemented to a glass slide and attached to an electric thermometer is a more sophisticated alternative. It should be noted that the temperature of a hot surface is not easy to measure accurately, nor is the temperature necessarily uniform over the whole area. With the thermoplastic resins, temperature is not critical and is best judged by experience. With at least one epoxy resin, temperature is an important factor controlling the refractive index of the hardened cement and very careful control is required.

Canada balsam

This is the traditional cement used by slide makers for more than 100 years, and still used by some makers for attaching the cover slip to the finished thin section. It is an exudation from the Balsam fir tree of Eastern Canada (*Abies balsamea*), which was discovered around 1680–1700 by a missionary sent to Canada by an ardent plant collector—Henry Compton, Bishop of London. Canada balsam has long been regarded as the finest cement for optical use and was almost certainly used by the lapidaries of the eighteenth century. Sorby used it to make his thin sections and it has been used ever since, though its popularity has waned somewhat with the introduction of epoxy resins, which give better adhesion to glass.

This resin is a pale brown, sweet smelling, very viscous liquid which sets hard when gently heated and cooled. It will melt when heated again and then re-harden and this process can be repeated over and over again, as long as the temperature does not become too high. If heated too much, Canada balsam darkens in colour, becomes brittle, and will not soften when reheated. In normal use it forms a clear and colourless film which does not darken with age, as is shown by Sorby's slides which show no sign of deterioration even after 140 years.

The other characteristic of Canada balsam which is particularly significant is its refractive index of approximately 1.54. This is close enough to that of many common rock-forming minerals to give good relief without being excessive.

To mount a slice with Canada balsam, the microscope slide and the rock slice are placed on a hotplate to warm. Using a glass rod (225 mm (9″) long and 6 mm ($\frac{1}{4}$″) diameter) a small amount of Canada balsam is transferred to the glass slide. The precise amount is a matter of experience, but as a rough guide a small puddle about 10 mm ($\frac{1}{2}$″) across and 2 mm ($\frac{1}{16}$″) thick is probably sufficient for a slice roughly 20 mm ($\frac{7}{8}$") square. Too much and the liquid balsam will run off the slide over the hotplate; too little and there will be insufficient to cover the area of the slice. The glass rod should not be returned to the bottle of balsam since it will gather up too much balsam for the next slide. It is best placed on a tin lid (or a plastic carton) on the bench away from the hotplate. If, by the time the rod has been disposed of, the balsam is smoking, the plate is too hot and the heat must be reduced. The first step is to 'cook' the balsam slowly, before mounting the slice. The balsam is tested by drawing out a strand between the points of a pair of forceps. When the strand snaps in the middle, the balsam is sufficiently cooked. Should the strand of balsam crumble, the balsam has been over-heated and the process must be repeated with a fresh slide and fresh balsam. This cooking process should take about 2–3 minutes or possibly a little longer. If cooking is not proceeding quickly enough, the heat can be increased slightly, but care must be taken not to heat the balsam too rapidly. The rock slice is then picked up with the forceps and carefully lowered on to the balsam, starting with an edge of the slice, in order to drive out air bubbles. Sometimes, bubbles occur in the cooked balsam, and these can be removed by bringing a heated needle close to them to burst them. If large numbers of small bubbles are present, this is usually a sign that the hotplate was too hot and that cooking has been too fast, even though the test with the forceps appeared satisfactory. Once the slice has been lowered on to the balsam, the glass slide and the slice should be removed to the bench and the slice pressed firmly down on to the glass. Careful examination of the underside of the glass should show a uniform film of Canada balsam with no sign of air bubbles or filamentous threads of air which often have a silvery appearance. The slide should then be left to cool slowly on the bench. If cooling is hastened, the balsam may develop cracks which will lead to the disintegration of the slice during later grinding. If air has been trapped between the slice and the glass, the slice can be removed by gently reheating and sliding it off the glass. The process of mounting is then repeated but this time, a little balsam can be placed on the rock slice and gently cooked, testing as before. Great care must be taken that this is done slowly to avoid overcooking the balsam already present.

Sorby, in his description of his technique in 1882, recommended the application of balsam to the slice as well as to the glass and scraping the excess of cooked balsam from the slice with the edge of a slide just before mounting. This has the advantage of filling pore spaces in the rock, especially in some sandstones. Generally, however, this is unnecessary and direct mounting of the clean slice is satisfactory.

Although mounting slices with Canada balsam in the manner described may appear to be very much a 'hit and miss' affair, various attempts to standardize heating rates, times, and temperatures seem to be fraught with even more problems. A little care, practice, and patience in cooking the balsam will yield acceptable mounts very quickly. Moreover, the technique, once mastered, is readily repeated.

The essential feature of a mount made with Canada balsam is that, when cold, the balsam should be hard enough to resist indentation by a fingernail, but not brittle and certainly not discoloured. If it is still soft, a little more gentle heating will probably suffice, but care must be taken not to heat the balsam so much that bubbles appear between the glass and the slice.

When the hotplate has cooled, spilled Canada balsam can easily be removed by scraping with the edge of a microscope slide. If any balsam remains, it can be removed with a rag moistened with methylated spirits. Soap, hot water, and a nail brush will remove remaining traces.

Lakeside 70C cement

This cement is sold in the form of short sticks of pale yellowish-brown resin. Its use is not unlike that of Canada balsam, except that no cooking is required. In thin films it is clear and colourless and has a refractive index similar to that of Canada balsam.

This resin melts quickly at about 80 °C and flows freely at 140 °C. The glass slide and the slice are heated to about 80–90 °C, and the tip of the resin stick applied to the glass sufficiently long to melt off a small pool of resin. A little resin is also applied to the surface of the slice which is then mounted on to the glass, pressed down, and allowed to cool as before. If the resin is heated above 155 °C, inadequate adhesion and some bubbling may occur and if held at this temperature for any length of time, the resin may burn on to the glass or the slice. Since there is negligible expansion and contraction of the resin when melted and cooled, no harm will be done by rapid cooling of the slide on a cold metal plate.

Lakeside 70C is very safe to use, giving off no toxic or disagreeable fumes in either the solid or molten state and is claimed not to produce any allergic reactions. As far as ease of use and cleanliness is concerned, there is no doubt that Lakeside 70C is superior to Canada balsam for slide mounting.

4.1.2. Epoxy resins

Cold-setting epoxy resins have become familiar in recent years under a variety of trade names, of which Araldite, Bison 'Combi-Standard', and Plastic Padding Super Epoxy glue are well known, as adhesives for a whole variety of materials, including ceramics and ferrous metals. Other epoxy resins and hardeners are available, but apparently only in large quantities or with excessive 'minimum order' charges: they are not referred to further.

These two-part cements are supplied as adhesive and hardener in separate containers. They are usually marketed in small flexible tubes or syringes. The two components are mixed immediately before use, usually in equal parts, but the manufacturer's instructions should always be followed carefully. Special care should be taken not to get these resins on the skin or to breathe the vapours.

For mounting rock slices using the material contained in tubes or syringes, equal amounts of the cement and hardener are squeezed on to the glass slide and mixed with a small piece of wood such as a matchstick. These adhesives are very viscous and may be softened somewhat by placing the closed containers in hot water for a short time before using. The addition of a drop or two of acetone will thin the mix, but this may prolong the gelling time. Care must be taken to ensure complete mixing of the two components, but this must not be so vigorous that air bubbles are trapped, which are almost impossible to remove. Some epoxy cements are almost impossible to homogenize completely, which makes them unsuitable for grain mounting. However, for mounting slices, this lack of homogeneity appears to have little or no effect on the finished slide.

The prepared slice is lowered on to the cement, starting at one edge and pressed down to squeeze out as much cement as possible. A spring-loaded clothes peg is a useful device for holding the slice on to the glass during the period of setting. More sophisticated devices which exert considerable pressure are available commercially (Fig. 7) and are essential for mounting slices for grinding on semi-automatic machines. Although heat can be applied to hasten the hardening of the cement, it is usual and probably preferable to set the mounted slices aside for 24 hours in a warm room or over a radiator before commencing grinding, to allow the cement to develop its full strength. Excessive heat must be avoided since stresses set up during the hardening process may be relieved by the disintegration of the slice during grinding.

Once the epoxy resin has gelled, the excess can be removed by scraping with a razor blade or sharp knife and the slide cleaned with a cloth moistened with alcohol (ethanol) or acetone. Once the resin has set hard, it cannot readily be removed.

Epoxy resins are now extensively used and because they have a high shear strength, the slice is not readily pulled from the glass during grinding. They are quite colourless when hard, but some are slightly opalescent, even in thin films. This is of no consequence when the slice is mounted on to the glass slide, but is undesirable when mounting the cover slip and other methods of completing the slide must be adopted, as described later (p. 36).

The refractive index of these resins varies, but is generally slightly higher than that of Canada balsam. As with all mounting media, if comparison of refractive index is to be made between the minerals of the rock slice and the

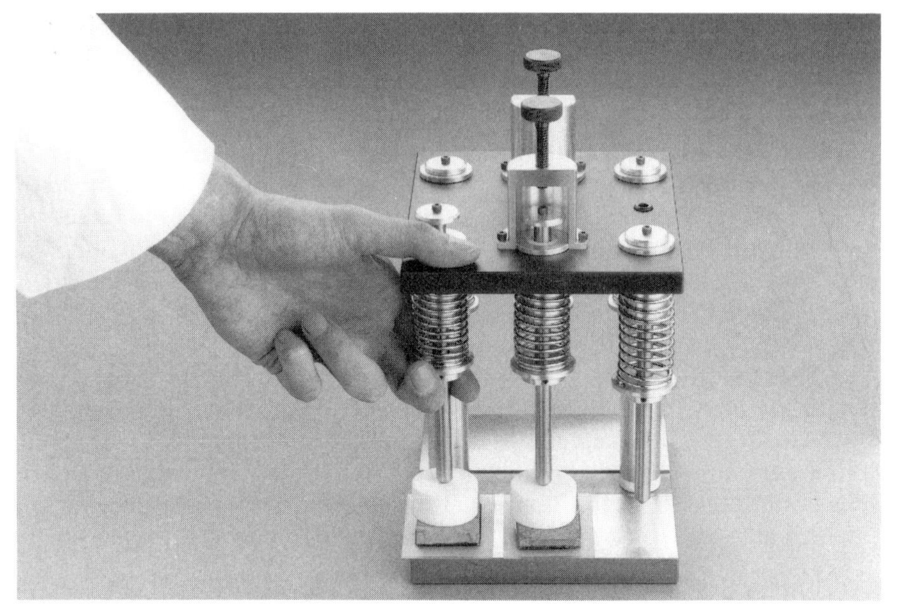

Fig. 7. Spring-loaded thin-section bonding jig. The jig can be stood on a hotplate to accelerate the setting of epoxy resins. (Photograph by courtesy of Logitech Ltd.)

cement, careful measurement of the refractive index of the cement must be made.

A recent introduction is the epoxy resin 'Petropoxy 154' manufactured by Palouse Petro Products, Washington, USA. It does, however, require careful heat treatment to cause setting and hardening, unlike the cold-setting types.

Among the advantages claimed for this resin are its refractive index of 1.540 ± 0.001 provided the manufacturer's instructions are strictly adhered to, its long pot life (up to 5 days at room temperature or indefinitely if refrigerated), its rapid curing (10 minutes), its very high shear strength, its low viscosity, and its clarity. The disadvantages appear to be relatively insignificant. Like all epoxides it may produce allergic reactions in some users if the exposure to the resin is considerable. Its strength may be sufficient under certain circumstances to break glass slides during the curing process because of shrinkage of the resin. A few minerals, such as halite, appear to inhibit the hardening of Petropoxy 154 and for these a cold-setting resin is preferable. This resin has been reported as attacking coccoliths and it is known to dissolve dyes such as alizarin, but the great majority of minerals and stains likely to be encountered in the study of rocks, minerals, and ceramics appear to be unaffected by it.

One other slight disadvantage of Petropoxy 154 concerns the problem of accurately dispensing the resin and curing agent, which are used in the ratio of 10:1, when only very small amounts of mixed resin are required. This is of some importance since the proportions of the two components markedly affects the refractive index of the hardened resin. Provided that a reasonable number of slides are to be made within a matter of a few weeks, the long pot life of the mixed resin, especially if kept refrigerated, should allow sufficient resin to be mixed with the necessary accuracy.

The mounting technique advised by the manufacturer is briefly as follows. The rock slice is thoroughly dried on a hotplate (or in an oven) for several hours. The mixture of resin and curing agent is applied to the surface of the specimen which has been heated to between 135 and 140 °C. A clean glass slide is then applied, avoiding trapping air bubbles (as with all cements). The slice and slide are then turned over on to the hotplate and the slice is pressed firmly into position. The slide is kept at 135 °C for 10 minutes to cure. After about 4 minutes the excess resin can be removed with a razor blade or sharp knife. Finally, the slide is removed from the hotplate, but while still hot, cleaned with a rag moistened with alcohol or methylated spirits.

4.1.3. Superglues and photocements

In the introductory remarks to this chapter, brief reference was made to the superglues (cyanoacrylate) and photocements. Neither has been used to any extent in slide making, though there has been some experimentation with superglues. Undoubtedly they will fix rock slices to glass, but trouble has been experienced with the rock pulling away from the glass where grinding has been carried out on power-driven laps. For hand grinding, however, they appear to be quite satisfactory. A disadvantage is the need to clean the edges of the rock slice after mounting, to remove the unhardened cement. There is also, of course, the danger of getting the slide stuck to the fingers or getting the fingers stuck to each other.

Photocements, such as Loctite Glass Bond, Bison Car Window Adhesive or Norland Optical Adhesive 61 do not suffer from the inherent dangers of superglues, unless one is particularly careless. They will fix rock slices to glass, but their behaviour during grinding is uncertain. Their usefulness appears to be in the ease with which mounts of sand grains and small fossils can be made, and in fixing cover glasses.

4.2. The determination of the refractive indices of mounting media

Although most of the mounting media are assumed to have a refractive index of around 1.54, there are occasions when it is necessary to determine this more accurately.

Most mountants undergo slight inelastic strain during curing and when the hardened cement is fractured. This strain can sometimes be detected as a faint rim of birefringent cement round the edge of an isolated grain. The usual techniques of measuring refractive index using immersion oils or by observing a polished slab in a refractometer do not, therefore, give wholly accurate results and methods based on mineral grains mounted in the cured cement are preferred.

The method devised by Scott Cornelius, of Palouse Petro Products, depends on the fact that uniaxial minerals have two indices of refraction mutually perpendicular to each other and parallel to the principal vibration directions. If such mineral grains are immersed in a fluid having a refractive index between the two indices of the mineral, there will be some position at which the mineral will have the same refractive index as the immersion fluid. If this position can be determined, then the refractive index of the immersion fluid can be readily calculated. In the present instance, the immersion fluid is, of course, the cured mounting medium.

Suitable minerals for this method, with an appropriate range of refractive indices, include some nephelines (RI 1.526–1.546) and quartz (1.544–1.553). The clear variety of calcite known as Iceland spar can be used for refractive indices in the range of 1.566 to 1.658. Because of its good cleavage, this mineral breaks into small rhomb-shaped particles which tend to settle on to the rhomb face. These show the two directions of vibration bisecting the angles of the rhomb, one direction giving ω (=1.658) and the other (shorter) diagonal giving ε' (=1.566). Thus a refractive index range of 1.526 to 1.658 with a small gap between 1.553 and 1.566 can be covered with just three minerals.

The appropriate mineral is crushed to a size suitable for microscopical examination (about 0.005–0.1 mm diameter) and part is mounted in the medium which is then cured under the conditions which will be used for all other mounting. The remainder of the mineral is used to determine the precise values of ε or ε' and ω by the standard method of using immersion liquids or by using mixed immersion liquids and a refractometer.

With nepheline or quartz, the slide is searched for individual grains giving a 'flash' figure when viewed in conoscopic conditions. In such grains, the *c* axis lies parallel to the plane of the microscope stage and the indicatrix is an ellipse whose semiaxes are ε and ω. With calcite, practically every grain will lie on its rhomb face and will show a high-order white colour between crossed polars.

With the lower polarizer in position, the angle of rotation (θ) during which the grain has a higher refractive index than the mounting medium (as determined by the Becke line test) is measured. For a complete revolution of 360°, θ will be measured twice and can be averaged. It is best to determine θ for a number of grains and average the results before making the calculation of the refractive index (η_D) from the following equation:

$$\eta_D = \left(\frac{\omega^2 \varepsilon^2}{\omega^2 \sin^2(\theta/2) + \varepsilon^2 \cos^2(\theta/2)}\right)^{1/2}.$$

For calcite, ε' replaces ε.

The above equation holds for optically negative minerals (e.g. nepheline and calcite). For positive minerals (e.g. quartz) the positions of ω and ε are interchanged.

This equation is readily derived from the equation for an ellipse by using the values of ω and ε or ε' for the semiaxes and by calculating the distance from the point on the ellipse given by θ to the centre (Appendix 2).

For the most accurate results, illumination of the microscope should be with a sodium lamp or with a suitable filter placed in the light path, such as Roscolux #23 Orange (Delly 1988).

4.3. Some notes on the safe handling of mounting media

Whereas Canada balsam and Lakeside 70, when used in the manner described, are non-toxic, the epoxy resins and photocements currently in use as mounting media may cause sensitization by skin contact or be irritants to the eyes, nose, and throat. This does not, however, imply that they cannot be used quite safely, provided sensible precautions are taken.

The safe handling of these mounting media (or any other chemicals) depends essentially on clean working, adequate ventilation, and not mixing substances if the outcome cannot be predicted.

Most of the mounting media recommended here are available in small flexible tubes with screw-on caps. Most of the epoxy resins are fairly viscous and some pressure is needed to exude them from the tube. Photocements, on the other hand, are much more fluid and very little pressure is required. Considerable care must be exercised when piercing the nozzle of tubes of photocements. With either type, the nozzle of the tube should be placed in close contact with the surface on which the material is required, the tube squeezed gently, and then lifted away carefully to avoid the formation of a long 'string' or thread of resin component. If a string forms, it can usually be broken by rotating the tube between the fingers or gently wiping the nozzle on the surface close to the original patch of material. On no account should it be broken with the fingers. The cap should be replaced on the tube immediately to prevent drops of resin falling on to the work surface. The makers' recommendation should be closely followed at all times.

Where resins are supplied in plastic bottles or wide-necked jars from which the contents have to be poured, some difficulty will be experienced in preventing liquid running down the outside of the container as pouring ceases. A crumpled tissue can be used to clean the container, but there is a serious risk of contaminating the fingers. A much better method is to use a 'spill' made from fairly stiff blotting paper to catch the last drop of resin. This spill is made from a rectangular piece of blotting paper about 150 × 75 mm (6″ × 3″), preferably torn, not cut, from a larger sheet so that it has slightly ragged edges. This rectangle is first folded into a square, then folded in half at right angles to the initial fold and finally folded again parallel to this second fold. This makes a stiff implement which can be handled at the folded end and easily rotated in the fingers to break any string of resin. It can also be used to wipe around the edge of the container and prevent the resin running down the side. The slightly ragged torn edge is more absorbent than a cut edge. A bundle of these spills should be made before work starts and stored in a small beaker. They can finally be disposed of by placing in an opened, heavy-duty polythene bag, which itself can be disposed of by sealing and sending for burial or incineration.

Containers used for mixing resins should also be disposable and no attempt should be made to clean them after use. The working area should be covered with aluminium foil or paper to facilitate the easy removal of drips or spillages.

Where small numbers of slides are being made and only small amounts of mounting media are being used, it is probably sufficient that a window is opened to provide some ventilation. On the other hand, where larger amounts of resin are being handled, all operations should be carried out in a well-lit fume cupboard fitted with an efficient extractor fan.

To avoid ingestion of any of these materials, eating, drinking, and smoking should be avoided when handling resins or photocements.

Should the user become contaminated with these materials, soap and cold water is the best way to remove them. Very viscous resins can be wiped off with a disposable cloth moistened with alcohol diluted with water, but other solvents should not be used, since these are often more dangerous to health and more readily absorbed into the skin than the resins which they are intended to remove.

It is often supposed that the problems of contamination can be avoided by the use of gloves and this practice is frequently recommended. There are, however, a number of objections to their use. Gloves are not totally impermeable to resins and after prolonged use, the operator may become contaminated without realizing that this has happened. In the author's opinion, which is based on his experience of handling much more toxic materials, there is a danger that the use of gloves makes the user careless and that he or she will tend to ignore contamination of the gloves with every likelihood that the resin will be spread to the outsides of tubes and bottles, to slides, to taps, and even to telephones. Unless the vapours of a substance are liable to be absorbed through the skin, when full protective clothing and a respirator are required, it is far better to be a clean and not a 'messy' worker and forego the use of gloves.

The inhalation of dust formed by sawing or dry grinding hardened epoxy resins can be avoided by wearing a face mask (which is advisable for all dry grinding operations) and wiping all working surfaces with a disposable damp cloth.

The hands should always be thoroughly washed after handling any of the mounting media. Overalls or a dust coat should be worn and should be laundered at frequent intervals to avoid contamination of clothing.

4.4. Summary

A brief summary of the characteristics and uses of the several mounting agents described in this chapter follows.

Canada balsam

- The traditional mounting agent.
- Viscous liquid, transparent.
- Thermoplastic—requires careful 'cooking'.
- Fairly easily cleaned from finished slide with alcohol
- Can be 'messy' to use if user is not careful.
- Used for mounting for hand grinding, and for hand-held grinding on laps—
  may not be strong enough for heavy grinding on milling type machines.
- Used for slide covering and for mounting loose grains.
- If correctly used, stable over very long periods of time.
- Refractive index = 1.54.
- Non-toxic.

Lakeside 70C

- A solid, thermoplastic resin—
  easier to use than Canada balsam.
- Stronger than but otherwise similar to Canada balsam.
- Long term stability probably good.
- Refractive index = 1.54.
- Non-toxic.

Epoxy resins—cold setting

- e.g. Araldite, Bison 'Combi-Standard', Plastic Padding 'Super Epoxy glue'.
- Two-part resins usually mixed in equal amounts prior to use.

- Moderate to high viscosity
- Sets hard at room temperature in about 24 hours—
  setting accelerated by heating.
- Short pot life.
- Transparent in very thin films—
  thick layers may be slightly opalescent.
- Very strong—can be used for all grinding.
- Clean in use—
  can be removed after gelling with sharp knife and alcohol—
  cannot be removed after setting.
- Not recommended for attaching cover glasses.
- Not recommended for mounting grains.
- Long term stability unknown.
- Refractive index ≃ 1.54–1.56.
- May cause sensitization by skin contact.

Epoxy resins—hot setting
- e.g. Petropoxy 154.
- Low viscosity—
  colourless and transparent.
- Two-part resin mixed in proportion of 10 parts resin to 1 part of curing agent—
  dispensing of correct proportions of components difficult in small quantities.
- Long pot life.
- Requires accurate temperature control for curing—
  curing temperature 135°C.
- Very strong bond, may break slide in some circumstances—
  can be removed after curing, while still hot, with a sharp knife and a rag dipped in alcohol.
- Suitable for attaching cover glasses.
- Suitable for mounting grains.
- Attacks coccoliths—
  influence on carbonates unknown.
- Long term stability unknown.
- Refractive index = 1.540 ± 0.001.
- May cause sensitization by skin contact.

Superglues
- Not recommended for slide making or for grain mounting.

Photocements
- Not recommended for mounting slices.
- Can be used for cover slips.

- Useful for grain mounting.
- Low viscosity.
- Clean to use.
- Harden quickly in daylight or artificial UV light—cannot be removed after hardening.
- Long term stability unknown.
- Refractive index ≃ 1.52.
- Irritant to skin and if inhaled.

5 Final grinding

Having mounted the rock slice on to its glass slide, the next step is to grind it to its final thickness. If heat has been used in the mounting process, the slide should be allowed to cool thoroughly before grinding commences. It should be inspected carefully to ensure that no air bubbles have been trapped between the glass and the rock slice and that the slice shows no evidence of lifting from the glass as a result of shrinkage of the cement and warping of either glass or slice. The occurrence of Newton's rings when the mounted slice is viewed in oblique light is a sure sign of an unsatisfactorily mounted specimen. Such slices should be removed from the glass, reground, and remounted.

5.1. Hand grinding

Grinding commences, as in the initial grinding with the coarse (120) grade carborundum and water, on a steel or glass plate. Using the whole area of the plate and with a circular movement, the carborundum is worked into a paste with the water and the rock gradually ground to a thickness of about 0.2 mm. At this thickness, most rocks are more or less translucent and some of the minerals will be almost transparent. However, it is essential that the greatest care has been taken to ensure that the top and bottom surfaces of the slice are as near parallel as possible before this thickness is attained. As an aid to this end, Sorby used to attach small squares of thin sheet zinc to the corners of his slides, with Canada balsam, when the slice was mounted. He was thus able to judge whether the slice was becoming wedge-like by observing the occurrence of abrasion on the surface of the metal squares. This method of control is rarely, if ever, used nowadays (probably because of the difficulty of obtaining the thin sheet zinc). A similar technique is sometimes used for small ceramic specimens which may be mounted within a thin slice cut from an alumina tube and which helps to prevent the development of a wedge section. However, usually it is possible to judge by eye and even by touch whether the slice is becoming wedge shaped. Often beginners at this art tend to put more weight on one end of the slide than the other and it is helpful if the slide is reversed end to end after every minute or so of grinding. Nevertheless, every attempt should be made to work with even pressure on both ends of the slide. In fact, great pressure is not required since this may cause the glass to bow which will either crack the slice away from the glass or produce a slice which is thick in the middle and thin at the ends (Fig. 8).

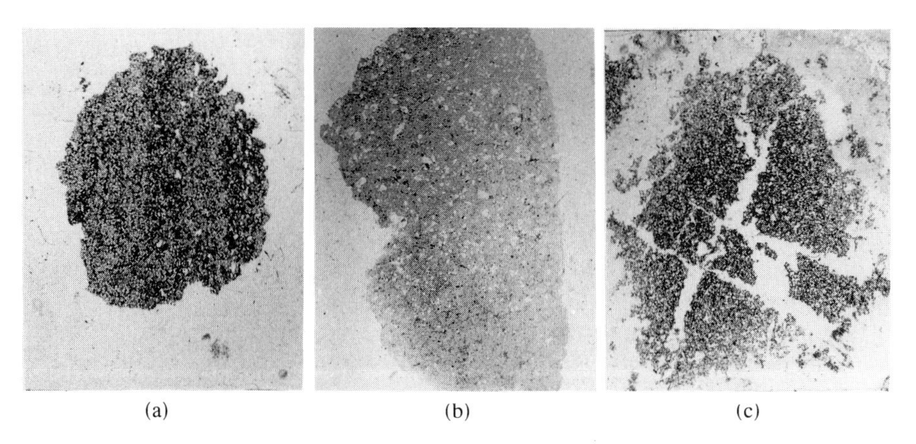

(a) (b) (c)

Fig. 8. Some faults in rock thin sections: (a) the first face of the rock was domed instead of being perfectly flat, with the result that the final grinding has produced a section that is nearly circular. (b) A wedge-shaped section resulting from excessive pressure along one edge in the final grinding. (c) Failure to use a fixative (see p. 36) has allowed the section to break up as the cover glass was lowered; magnification ×1.6.

Fresh carborundum powder can be added from time to time as grinding proceeds, together with a small amount of water so that a fairly thick slurry of carborundum, rock dust, and water is kept on the plate. However, carborundum should not be added when the slice is approaching its final thickness, in order to avoid the formation of deep scratches which will have to be removed with the next finer grade. This stage of grinding should be characterized by the almost complete absence of the distinctive scratching noise associated with new grit.

During this phase of grinding, the slide should be washed periodically. If a tap is used, care must be taken not to get any grit on to the tap handle which might subsequently be transferred to the next stage of grinding.

When the slice is at the correct and uniform thickness it is washed clean of all traces of carborundum and rock flour. A small nail-brush is helpful in dislodging the last grains and then the hands should receive similar attention. It is vital that no coarse grit is transferred to the next grinding plate, otherwise much hard work may be spoilt by the appearance of deep scratches in the nearly finished slide.

The next stage of grinding is with 220 grade grit on a glass plate to reduce the thickness to about 0.1 mm. At this thickness, quartz will show bright second-order interference colours when examined under the microscope with crossed polars. To reach this thickness (or thinness) requires considerable care, and attention must be paid to retaining the uniform thickness of the slice as judged by visual appearance. The circular motion used for grinding in earlier stages can still be used, but some straight-line movement up and down the plate, with a very light touch, is often advantageous. Inevitably, the ends of the glass slide will show evidence of abrasion, but this does not necessarily mean that the slide is becoming 'wedgy' or that it is being rocked from side to side, though this must clearly be avoided. It is associated with grit being trapped beneath the glass, especially if the paste becomes too dry. Judicious addition of fresh grit and water is necessary to keep the grinding 'sweet'. When the slice is about 0.1 mm thick as judged by the interference colours under the microscope, it is once more thoroughly cleaned, as are the hands. The slice is now ready for the final stage.

Grinding the slice down from 0.1 mm to the standard thickness of 30 μm requires considerable skill. Some acquire this skill very quickly with only one or two lost slides, others find it takes longer. The secret is in a firm, but gentle touch, with no 'scrubbing' of the slide in a small area of the plate.

The final stage of grinding is done with the 3F grade of grit on a glass plate and long, steady strokes, using the whole area of the plate as before. The mixture of grit and water should not be allowed to become too dilute by the addition of too much water or the slide may stick to the plate, nor should too much grit be added or the slide will again stick. It is best to mix the grit and water on the plate with the finger tips, both before and during grinding. To remove the slide from the plate, it should be slid off the edge of the plate; no attempt should be made to lift it off by pulling it upwards.

The slide should be examined frequently under crossed polars and grinding continued until the quartz grains show white or, at most, very slightly yellow, first-order interference tints. In these final stages a few gentle strokes up and down or across the plate with the slide moved with the middle finger of one hand is often all that is required to complete the thin section.

If, as sometimes happens, the slide suddenly begins to break up, the best thing to do is to put the slide on one side for cleaning and covering. Sadly, this is also the stage at which the rock slice may disappear altogether and the only solution is to start all over again. Usually, this happens when the operator's attention is distracted or he or she becomes impatient and rubs the slide a little too hard.

It has been assumed that quartz is present in the rocks being sectioned and that the thickness of the slice will be judged on the appearance of that mineral. It is true that many rocks do not contain quartz, but it is preferable to master the art of slide making using quartz-bearing rocks. Feldspars can also be used as thickness indicators and will show first-order grey or white colours at the standard thickness. The use of other minerals will depend on their abundance in the rock, the ability of the slide maker to identify them at an early stage of grinding, and to use an interference chart to judge the slice thickness.

5.2. Machine grinding

The early part of the final grinding can, with advantage, be carried out on the simple grinding wheels described in Chapter 3. The procedure is similar to that described above for hand grinding, with the reminder that great care must be taken not to allow the slide to be snatched from the fingers and thrown against the wall of the machine. If this happens the chances of springing the slice from the slide or breaking the slide are considerable. To minimize this danger, it is advisable to run the wheels at a low speed (20–30 r.p.m.), if that is possible. An ample supply of grit and water added with a brush from time to time also reduces the likelihood of the slide sticking to the wheel. Slides are finished by hand as described above.

As noted earlier, there has been a move towards automation of slide making with several slides ground and finished simultaneously in purpose made chucks (Fig. 6). The mounted slices are held in the chuck either by vacuum or by a thin film of wax. As in the initial grinding, the whole process is carried out using the finest carborundum grit (600 mesh or 3F). Since the slides cannot be easily removed from the chuck for checking their thickness optically, reliance has to be placed on readings of a dial micrometer attached to the mounting plate or the chuck adjusted so that the slides begin to 'float' when the correct thickness is reached. Much reliance is placed on the uniformity of the thickness of the glass slides on which the rock slices are mounted and it is generally necessary to pick slides of uniform thickness (measured by micrometer) from a box full of slides or to grind sets of slides to a uniform thickness before mounting the slices.

Power grinding on the larger, milling-type grinding machine with adequate control of the position of the grinding table allows grinding to proceed much further than with grinding by hand or by the simple machines. The main difficulty is knowing when to stop, since it is almost impossible to remove the specimen from this type of machine for examination under the microscope and then to replace it in exactly the same position. However, with suitably designed machines it is possible to control the grinding to a thickness such that a minimum of hand finishing is required. This presupposes that the grit size of the wheel is sufficiently fine for a finishing cut to be taken, which in turn means that the early stages of grinding are likely to be somewhat prolonged. This can be overcome to some extent by taking a deep cut and using a slow traversing speed, with an adequate supply of water as lubricant and coolant. In such circumstances, the use of the strong epoxy-type resins is essential to avoid lifting the slice from the slide.

6 Transferring and covering

The final step in making a thin section is to transfer it from the slide on which it was made, if that is possible, and to cover it to protect it from damage.

Traditionally, when Canada balsam was the only cement used for slide making, almost all sections were transferred to clean microscope slides. Slices mounted with Lakeside cement can also be transferred, but those mounted with thermo-setting epoxy resins and the photocements cannot be removed from the glass slide upon which they were made. The abrasions on the ends of the glass slide which inevitably occur therefore need to be covered with paper labels. It is sometimes suggested that the whole surface of the slide is ground with the finest grade of carborundum before the rock slice is mounted. This does make for a neater slide on completion, though even this surface may become scratched in the early stages of grinding. On the whole it is probably best omitted.

6.1. Transferring and covering the slice

Before the slice can be transferred, it usually needs to be strengthened to prevent it breaking up during manipulation. This is achieved by lightly coating the surface of the slice, but not the surrounding cement, with a thin layer of a suitable fixative. This used to be a dilute solution of pyroxylin in amyl acetate, but nowadays a cellulose lacquer (or a colourless nail-varnish) has largely replaced it. The fixative is painted on to the slice with a small camel-hair brush and dried quickly by gently blowing over it. Spray-cans of cellulose lacquer can be used, but the area sprayed is not easily controlled and, furthermore, bubbles of the propellant gas may form in the layer of lacquer and may not burst before the lacquer is dry.

Excess cement around the slice is removed with a sharp knife or razor blade, care being taken not to lift the slice from the glass. The slide can be cleaned with a cloth dampened with alcohol (methylated spirits).

The mounted slice and a clean microscope slide are placed on the hotplate which should be at about 70–80 °C, and a small amount of Canada balsam or Lakeside 70C is placed on each. The temperature of the hotplate should be allowed to rise until the added cement is completely liquid and then the cover slip is gently lowered over the rock slice. Gentle pressure on the edge of the cover slip with a match stick or similar peg sharpened to a chisel edge will detach the slice from the glass and cause it to rise up under the cover slip. Without lifting the cover slip or the attached slice, they are gently pushed across on to the new slide, adjusted into position, and very carefully pressed down to expel as much of the cement as possible. If the slice is out of position, it can be moved by squeezing down on one edge of the cover slip with the wooden peg. Great care must be taken not to press too hard and thus break the cover slip. Should this happen, the broken fragments can generally be removed, a little more balsam added, and a new cover slip applied. The temperature of the hotplate should be allowed to rise slightly (to about 90 °C) and the slide left for about 1 minute before removing it to a cooler part of the plate. Finally, it is removed to a wooden block to cool slowly to room temperature. It will be noted that the balsam is deliberately 'undercooked'; nevertheless it will set and the life of the slide is much prolonged.

Air bubbles sometimes occur under the cover slip. These may be due to the entrapment of air as the slip is lowered into the balsam and can usually be squeezed out by judicious pressure on the surface of the slip, though if the slip is lowered carefully from one edge, air bubbles should not be trapped. Bubbles may also occur after the slip is in position. These are almost certainly due to overheating the balsam and the remedy is simply to avoid excessive heat in the covering procedure.

When mounting with Lakeside 70C, no 'cooking' is involved and the slide can be removed as soon as the cover glass is in place, the slice centred, and the slip pressed well down. When the slide has completely cooled the excess cement is very carefully scraped away with a sharp knife or razor blade, again taking care not to spring the cover slip from the slide. Final cleaning with a cloth and alcohol should complete the production of the thin section. Sometimes, a gentle scrubbing in hot, soapy water is needed to give a clean slide, but generally this should not be necessary.

The cover glasses (slips) used to cover rock and mineral sections are always square or rectangular (never circular) and usually 22 × 22 mm ($\frac{7}{8}$″ × $\frac{7}{8}$″) or 22 × 25 mm ($\frac{7}{8}$″ × 1″). Very thin (No 1) cover slips are rarely used since high magnifications are not often required in petrology. No 2 or No 3 slips are much less fragile and are usually quite satisfactory.

6.2. Covering without transfer

As noted above, slides made with epoxy resin or photocement mountants cannot be transferred. Nor is it always necessary to transfer sections mounted with Canada balsam or Lakeside 70C. Three options are then open: hot covering, cold covering, and lacquering. These procedures are described below.

1. **Hot covering.** Slides can be covered with a glass slip affixed with Canada balsam or Lakeside 70C in a manner similar to that described

above. Slides mounted with Canada balsam or Lakeside 70C will require treatment with a fixative, but those mounted with epoxy resins or photocements will not, since there is no chance of the section breaking up. The original mountant is cleaned from the edges of the slide as before. The slide is then placed on the hotplate at about 80 °C, and the slice covered with Canada balsam or Lakeside 70C. The cover slip is gently lowered on to the slide and pressed down. The slide is cooled and cleaned as described earlier.

Alternatively, a hot-setting epoxy resin can be used on slides originally mounted with epoxy resins or photocements. It should not be used on slides mounted with Canada balsam or Lakeside 70C. The temperature and time of heating should follow the manufacturer's recommendation.

2. **Cold covering.** Cold-setting epoxy resins, photocements, and superglues can also be used for attaching the cover slip with the advantage that no heating is needed, though bright sunlight or a UV lamp is required to cause the photocements to harden. The chief objection to the epoxy cements of the two-tube DIY variety is their high viscosity, making it difficult to press down the cover slip sufficiently without breaking it. The superglues do not produce a very strong bond between glass and rock section, but it is probably strong enough for the present purpose. Of the three methods of cold covering, the photocements appear to be the most suitable.

3. **Lacquering.** The use of cellulose lacquer, applied either by brush or as a spray, has been recommended as a means of protecting the finished thin section. However, such a covering is liable to become scratched and certainly cannot be regarded as a permanent means of preserving a thin section on which a great deal of time and effort has been expended.

If thin sections are to be stained or examined by cathodoluminescence, they should not, of course, be covered.

7 Labelling and storage

It should be unnecessary to remind the reader that the finished thin section should be labelled. Some indication of the source of the section should be given, so that at a later date confusion cannot arise. If a name is to be given, it can be added on completion of the microscopical examination of the slide.

The usual method is to attach gummed paper labels and to use a permanent ink, preferably Indian ink. Unfortunately, paper labels tend to come unstuck with the passing of time and they also get dirty and become illegible. The only solution to this is to go back to the method adopted by Sorby and to use a writing diamond directly on the glass. This may seem to be a counsel of perfection, but Sorby's slides made 140 years ago still have legible legends. But it is unlikely that many microscopists will have the confidence to believe that their slides will still be worth looking at 140 years hence, and paper labels will, no doubt, continue to be used.

The finished slides should be stored in slide boxes with vertical slots (which are economical in space) or on flat cardboard trays (where they are readily visible). They should not be kept in a heap in a specimen tray or left lying in the bottom of the microscope case where cover slips can be knocked off or get broken.

The cataloguing of collections of thin sections is a topic beyond the scope of the present text, but some thought should be given to it when slides are labelled and storage is considered.

Fig. 9. One of the tin boxes used by H. C. Sorby to contain his slide collection.

8 Special methods

So far, emphasis has been on the procedures for the production of a thin section which, if the suggestions and instructions given earlier have been followed, is likely to have been straightforward. However, there will undoubtedly be instances where the methods given are not immediately successful, but it is to be hoped that on the basis of the principles given earlier and in this chapter, the reader will be able to devise a solution to his or her particular problem.

For example, the material may be too friable to slice, or too thin and brittle, such as an egg shell. The sample may be too small to handle easily or it may consist of hard grains in a soft matrix. The material may be soluble in water or sensitive to water (anhydrous materials). It may be desirable to stain the components of a rock in order to facilitate the discrimination of the minerals present. For some purposes, very thin sections only 10 μm thick may be required. Sometimes, the preparation of a thin section may not even be necessary if interest centres on the texture of the material, rather than its composition.

8.1. Consolidation of friable materials

A commonly encountered problem is that of obtaining a suitable slice of a friable material for initial grinding. Some sandstones and the soft chalk of southern England are characteristic of this group of materials. They are often too soft to be clamped firmly for machine-sawing and the only method of reducing a specimen to a suitable size is to flake a piece off with a small square-paned geological hammer. The flake is unlikely to have a flat face, but gentle abrasion on a sheet of fine carborundum paper, used dry, will produce a suitable surface, but it will usually produce a lot of dust which must be blown away. This dust should not be washed off, since washing may push the dust deep into the fragment. With care a second face parallel to the first may be formed on the specimen to give it some stability for the next step. The specimen is placed on a hotplate with the larger face uppermost and warmed through—to a temperature of about 60 °C. A little Canada balsam is then dropped on to the uppermost surface and as it soaks into the specimen more is added until 3 or 4 mm of the rock is saturated. It may be helpful to put a few drops of pure turpentine (not turpentine substitute) on the rock surface before adding the balsam. The rock is then kept hot, but not too hot, until the balsam has hardened. The specimen can then be removed and allowed to cool slowly. From this point, the specimen can be ground and mounted on a glass slide and finished as previously described.

An alternative method of impregnating friable or very porous rocks is by total immersion in a mixture of Canada balsam and shellac. The method is to heat a sufficient quantity of balsam in a small crucible over the flame of a Bunsen burner until the balsam boils very gently. As the balsam is heating, about 15 per cent by volume of flaked shellac is dissolved in the balsam. The cold dry specimen is then immersed in the mixture. There is usually vigorous, or even violent boiling as air escapes from the pore spaces. The action gradually subsides as the medium penetrates and in about ten or fifteen minutes the balsam will boil gently when impregnation is complete. The specimen is removed from the still-hot liquid, allowed to cool, and the thin section made in the usual way.

The above methods will probably be regarded as 'old-fashioned' by some, but they do have the virtue of requiring very little apparatus, though they may perhaps be rather extravagant in the use of Canada balsam.

Most impregnation is done at the present time with epoxy resins, such as 'Epo-tek', a cold-setting resin, or with 'Araldite' diluted with toluene. It is insufficient merely to immerse the rock fragment in the resin because despite the low viscosity of some resins, they will not penetrate the smallest pore spaces unaided. The simplest method is to place the specimen in a small disposable pot or a mould made from folded aluminium foil and cover it with the resin. The pot (or mould) and resin are placed in a desiccator attached to a water pump and the pressure lowered as far as possible. Care should be taken to include a trap in the vacuum line to prevent water reaching the desiccator should the water pressure drop. When gassing of the specimen and resin ceases, the air pressure can be slowly raised and the pot set aside for the resin to harden completely. An alternative method is to outgas the specimen before adding the resin by using a tap funnel through the top of the desiccator. This has the disadvantage that the resin itself is not outgassed before it enters the pores of the rock. To overcome this problem Miller (in Tucker 1988) has devised a chamber with rotating spindles carrying narrow blades which can support the rock fragment over the resin while both are outgassed. Rotation of the spindle then drops the fragment into the resin.

Resins suitable for impregnation include Araldite AY-18 which has low viscosity which is retained for up to a week after mixing. Hardening is initiated by raising the temperature of the resin to 80 °C. Other resins, such as 'Epofix' and Trylon CL 223 PA can also be used and there is considerable scope for experimentation in this field.

To emphasize the pores and channels in rocks, it is helpful if the epoxy resin is stained with a blue dyestuff, such as Keystone Oil Blue or Waxoline Blue, in the ratio of about 10–15 per cent dyestuff to resin.

8.2. Embedding specimens

Small fragments of material, such as drill chippings, small shells, or loose sand grains cannot easily be sectioned by the methods already described without some preliminary treatment to allow them to be more readily handled. Probably the simplest method is to cast the fragment or the grains into a mass of resin contained in a small mould about 12 mm ($\frac{1}{2}$") square and sufficiently deep for the resin to cover the specimen. When the resin has completely cured, the specimen can be manipulated with ease and can be seen through the resin. A silicone releasing agent, such as a domestic silicone-based furniture polish in a spray-can, applied to the mould before casting, facilitates the removal of the block of resin. It is helpful if the sides of the mould slope outwards at a slight angle.

An older method was to mount the specimen in plaster of Paris. The disadvantage is that it is, of course, opaque and the position of the specimen cannot be seen until after grinding has commenced. It is also very soft and needs to be impregnated with balsam and shellac. There is now little to recommend this method.

8.3. Sectioning hard minerals in a soft matrix

The occurrence of hard mineral grains in a soft matrix is not an uncommon one. Weathered rocks and some ceramics are typical of this type of material, as are sandy clays. The matrix may consist of large grains of soft material or it may be a mass of very fine particles. In either case, the main problem that arises is the pulling out of the harder grains from the matrix or the more rapid removal of the soft matrix, in the initial grinding, making it difficult, if not impossible, to obtain a truly flat surface. Should an attempt be made to finish such slides, the matrix is likely to disappear completely before the hard grains have reached the desired thickness.

In all materials of this type impregnation is the essential initial step using one of the methods indicated in Section 8.1. In passing, it may be noted that many materials requiring this treatment can be cut with a fine-toothed hacksaw blade, though care must be taken not to crush the material when it is clamped for sawing.

For most materials of the type being considered here, the specimen should be ground using only the very fine 600 grade carborundum. For hand grinding, this does mean considerable labour, but it is preferable to losing the slice. Machine grinding with the specimen held in a chuck rather than in the hand gives the most satisfactory results.

Should the hard component of a rock or ceramic be as hard or harder than carborundum then it will be necessary to use tungsten carbide or diamond as the abrasive.

8.4. Sectioning of 'water-sensitive' materials

Though many naturally occurring materials are soluble in water to some extent, the solubility is generally so small that the use of water in section making has no visible effect on the finished thin section. There are, however, a few minerals such as the evaporite minerals which are very soluble in water (e.g. halite or rock salt) or which react with water (e.g. anhydrite) that cannot be sectioned by the methods already outlined.

Another group of minerals which should probably be included here are those which oxidize fairly rapidly in the atmosphere. These include the sedimentary iron minerals, chamosite and siderite. Glauconite should probably be treated as unstable, but only in exceptional circumstances of high abundance and extreme freshness in a sediment does it receive special treatment.

There are two solutions to this problem. The first is to use a non-aqueous carrier for the abrasive; the second is 'dry grinding', described in the next section.

The traditional liquid for this purpose was medicinal or liquid paraffin. Subsequently, light mineral oil, kerosene, or glycol (ethylene glycol) was used in place of water and the grinding process exactly paralleled that with water. The main difficulty with using these liquids was that of cleaning the slides, the grinding plates, and the machines. Liquid paraffin can be removed with 'white spirit' (naphtha); kerosene can be used to wash off kerosene and mineral oil, while glycol can be removed with alcohol. Adequate health and fire precautions must be observed in handling these materials. Glycol and alcohol are, perhaps, the least offensive in this respect.

In most cases excessive heat should not be used at any stage in the process. Anhydrite will decrepitate, gypsum will dehydrate, halite is fairly stable, and chamosite will recrystallize to a flaky mass of yellow and brown material, not unlike weathering products. Cold mounting and cold covering are therefore preferred.

8.5. Dry grinding

The dry grinding of rock, mineral, and ceramic materials is a comparatively recent development, though it has been used by metallographers for many years. Although this procedure is more suited to machine grinding, it can readily be applied to hand grinding of soft materials, such as clay and chalk and to the water-sensitive materials referred to in Section 8.4.

The simplest form of dry grinding is to use a fine-cut file to produce a flat surface on the specimen. This is followed by grinding the specimen on a series of carborundum papers stuck down on to flat pieces of wood. The papers are commonly sold in DIY stores as silicon carbide papers or 'wet-and-dry' carborundum papers in a variety of grades. The finest grade commonly available appears to be 400 mesh. Finer grades up to 2000 mesh are made for polishing metals and ceramics.

With care, thin sections of soft materials can be made using the finest grades of paper without the need for impregnation. The process is similar to wet grinding, but care must be taken that grinding is not so vigorous that sufficient heat is generated to damage the specimen or melt a thermoplastic adhesive. It is therefore advisable to grind the specimen for 30–40 seconds and then allow it to cool. The dust produced in grinding should be blown off rather than washed off.

To avoid contamination, grinding sheets should be kept covered when not in use.

8.6. Two methods of making 'pseudo-sections' of clays and shales

Thin sections of clays and shales are notoriously difficult to make sufficiently thin to reveal very much of the mineralogy of the material. Usually only the coarsest grains are sectioned and appear to occur in a fuzzy, indeterminate groundmass.

One method of overcoming this difficulty was devised by Weatherhead (1947). Instead of attempting to make thin sections of clays, he made smears on glass slides which were prepared by grinding the slides on 220 grade carborundum on glass plates. These slides were then rubbed face to face to remove glass shards and thoroughly washed and dried. The slightly moistened clay specimen was then rubbed on the ground surface to leave a smear of particles. Excess material was removed by gentle brushing. These could be examined as temporary mounts under glycerine or coated with balsam and permanently mounted. Any structure within the specimen is, of course, destroyed by this method.

Another method which may work in some instances, though it has not been followed up, is based on a chance observation of the author's many years ago. A technician, making a slide of a shale, left the mounted slice in water while he went for his lunch. On his return, the section had apparently disappeared from the slide. In fact, a very thin layer of shale, only a grain or two thick, remained adhering to the glass. The adhesive was Canada balsam, but there is no reason to suppose that other adhesives would not have behaved in the same way, penetrating very slightly into the specimen and fixing it to the slide. Not all clays or shales will behave in this manner; much depends upon the amount and nature of the cementing agent within the rock.

Probably this technique will work best with uncemented clays which should be ground dry to a smooth face before mounting and then immersed in water. Gentle scrubbing may accelerate the breakdown of the rock. Ultrasonic disintegration might be effective provided it does not break the bond between the adhesive and the glass.

8.7. Mounting mineral grains and fibres

The procedure for mounting mineral grains and fibres is similar to that already described for mounting rock slices. Any of the mounting agents used for rocks can be used for grains and fibres.

Traditionally, mineral grains were mounted in Canada balsam. The method is to place a small amount of balsam on a glass slide and gently cook it until a thin strand drawn off with the points of forceps just breaks as the points are opened. The grains are then placed on the balsam and stirred into the centre of the pool of liquid. As far as is possible, they should not be dispersed. The edge of a warmed cover slip is then laid alongside the balsam and carefully lowered. The cover glass is gently pressed down to spread the balsam and this, in turn, speads the grains. If the quantity of grains added initially is correct, no grains should escape from under the slip. Only experience will show the correct amount, but enough to cover the tip of a penknife is a reasonable starting point. Fibres are, of course, treated in the same way. One problem that may arise is the formation of gas bubbles in the balsam as it is cooked, due to rapid or prolonged heating. These can be removed by pricking them with a needle or wafting a micro-gas jet over the surface. The slide can be cleaned by scraping and wiping with methylated spirit as for rock slices.

Lakeside 70 resin is a little more difficult to use, since it less readily forms a globule when melted on to the slide. However, raising the temperature of the slide to about 90 °C will usually give a suitable pool of adhesive.

Photocements can be used in a similar way, though the excess cement must be removed before exposure to ultraviolet light. It is also possible to use epoxy resins to mount grains, but in general, they are not recommended since they are commonly slightly opalescent and are not optically homogeneous in the thick layers required. With a little practice, the amount of adhesive used can be such that very little cleaning is necessary, but for a neat slide, it is essential that sufficient is used to reach the corners of the cover slip.

When a grain mount is examined under low-power magnification the grains should be fairly evenly spaced with little more than a grain diameter between each grain. They should not be touching and they should not all be along one edge of the cover slip. A gentle circular motion of the cover slip as it is pressed down will help to spread the grains more evenly. Ideally, every grain should be resting on its largest face, but if a few grains much larger than the remainder are present, these will prevent the bulk of the grains attaining the desired orientation. For the best slides, the grains should be as uniform in size as possible.

Small organisms such as foraminifera and diatoms could be mounted in this way, but it is more usual to handpick these and place them individually on a slide. Gum tragacanth is a common adhesive for fixing them, but a method using Norland Optical Adhesive Type 61 (a photocement) diluted with acetone has been devised by Porguen (1989). In brief, these small organisms are transferred to the slide with a fine brush moistened with the diluted photocement and fixed in position by exposing the slide to sunlight after the solvent has evaporated. The cover slip can then be fixed with the undiluted cement and hardened as before. Since the mounts become permanent only when exposed to ultraviolet light, working in the normal environment of tungsten lighting or away from direct sunlight allows considerable time for the preparation of individual mounts.

Porguen (1989) advises that slides should be thoroughly cleaned with hot detergent or 'sulphochromic mixture' (a potent cleaner, but highly dangerous). The author has found that photocements adhere perfectly well to glass without special cleaning other than a rub with a cloth or tissue.

8.8. 'Half pebbles'

The lithology of pebbles in sandstones is often a clue to the provenance of the rock. These pebbles can be sectioned in the normal way, but this is time-consuming and not always necessary. If the pebble is ground to a flat surface with fine carborundum and mounted on a glass slide, it can be examined, at low magnification *through* the slide with a binocular (stereo) microscope (Fig. 10). By their nature, pebbles are usually fairly tough rocks and are often highly siliceous. Consequently, the internal character can be

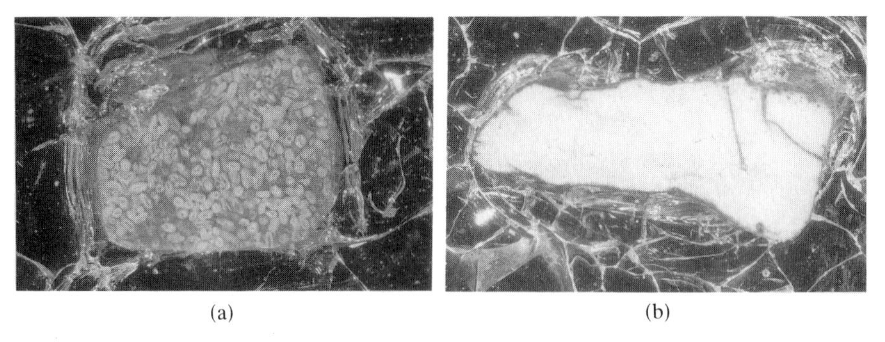

(a) (b)

Fig. 10. Two mounted 'half pebbles'. (a) A silicified oolitic limestone; (b) a featureless quartzite. Note that the Canada balsam was somewhat overheated when the pebbles were mounted and has developed small cracks; magnification ×6.

observed, not merely the ground surface. Many pebbles can be treated in this way very quickly; many will be seen immediately to be quartzite or chert: others may contain microfossils, which, if they can be identified, are valuable indicators of the source of the pebble, while others may be of special interest requiring a thin section. These can be finished as such, thus saving much labour by not sectioning all the pebbles.

8.9. Orientated mineral grains

For some crystallographic and teaching purposes it may be necessary to section crystals of minerals in particular directions. This requires some considerable skill and crystallographic knowledge on the part of the slide maker.

The main problem is orientating the slice with the required section parallel to the grinding surface. While special mounting devices do exist, the usual method, if the crystal is large enough, is to cut it manually on a diamond wheel in the required direction, or if too small to hold easily, to mount it in a resin block large enough to handle. The slice is then ground on each side and the orientation checked under the polarizing microscope. The orientation can be adjusted by grinding one edge more than the other, which takes considerable skill. When the required orientation has been achieved by, for example, centring a conoscopic figure, the section is mounted and finished by the usual methods.

It may be noted here that many small mineral grains, when mounted as described earlier (Section 8.8), take up a preferred orientation depending on their physical shape. Micas, for example, always mount with their basal cleavage parallel to the slide, while un-worn zircon grains will lie with their *c* axis parallel to the plane of the slide. Well abraded zircons, on the other hand, take up a random orientation.

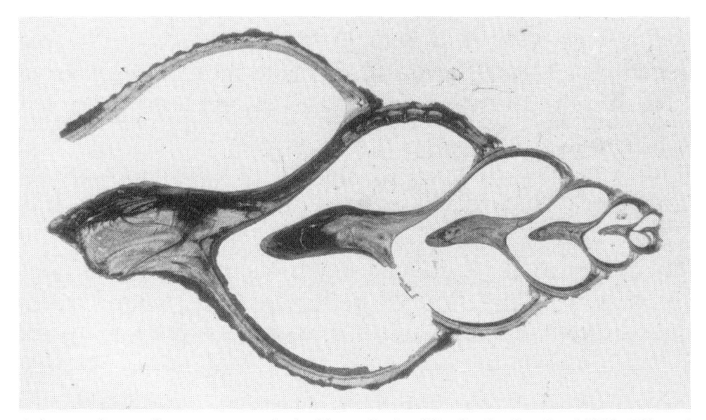

Fig. 11. A thin section of a gastropod shell made by H. C. Sorby in 1852. The shell is about 20 mm ($\frac{7}{8}$") long.

9 Polishing for reflected light microscopy

If a dry, uncovered thin section, prepared as described in previous chapters, is examined under the microscope by reflected light, it will be seen to be covered with minute scratches which severely degrade the image. The application of the cover slip effectively removes the scratches but introduces a reflecting surface which itself degrades the image (in reflected light). For reflected light microscopy it is therefore necessary to dispense with the cover slip and improve the surface finish of the section. This may be achieved by *polishing* the surface to remove the scratches and produce a near mirror-like finish.

The production of a polished surface on a specimen is an extension of the procedures already described, since polishing is simply an extension of grinding by removing smaller and smaller surface imperfections. Polishing is therefore a lengthy process which can be carried out by hand, but is more generally performed by machine methods. Furthermore, machine methods produce much flatter surfaces than can be prepared by hand.

There is, however, an important distinction between the techniques of grinding (or lapping) and polishing. Grinding uses a lap made of a material harder than the specimen being ground, whereas polishing uses a lap which is softer than the specimen. With the hard lap, the abrasive is held on the surface and strikes the surface being cut with considerable force and abrades it quickly, but with a soft lap, the abrasive can impregnate the lap and the shock of striking the specimen is cushioned to some extent. The fragments being removed by polishing tend therefore to be smaller than those produced by grinding, even if the same grade of abrasive is used. Generally, of course, very much finer abrasives are used for polishing than are used for grinding.

The technique of polishing can be applied to transparent or translucent materials, thereby producing thin sections of very high quality, and it can be used to produce ultra-thin (10 μm) sections. However, its most important use is in the preparation of opaque materials, such as ore minerals and many ceramics for examination by reflected light, where freedom from scratches and other surface imperfections is essential. Polishing also produces truly flat surfaces, free from relief, and is therefore a valuable technique for the preparation of those materials containing mixtures of hard and soft components.

The practice of polishing has been applied to thick slices of cement clinkers, although thin sections of these materials can be prepared. The polished surface is said to be more suitable for use with etching reagents than uncovered thin sections.

9.1. Polishing laps

A variety of materials can be used for polishing laps, some of which are described below.

- *Cast iron.* This is the commonest material used for grinding and with sufficiently fine grained abrasives will produce a low degree of polish on most specimens.
- *Soft metals.* These include copper, tin, lead/tin alloys, and pure lead, the latter being the most commonly used material. The soft metal is usually bonded to a cast iron or steel plate which can readily be attached to or removed from the driving spindle of the grinding/polishing machine.
- *Cloths.* These include both woven and non-woven materials made from a variety of textiles, including artificial silk, nylon, rayon, wool, and paper. They are usually supplied with a self-adhesive backing for fixing to a backing plate. An important characteristic of all cloth discs is their ability to withstand loading of the specimen and their ability to retain the abrasive in an efficient and economical manner.
- *Other materials.* Various other materials can be used for polishing including synthetic resins, wood, and the original polishing material, pitch. There are also composite materials in which metal powders including copper, iron, tin, and lead are bonded with a mixture of plastics. Marketed under the trade name *Kemet* they were developed by Engis Ltd of Maidstone, Kent. It is claimed that they can be matched to the hardness of the material being polished with the complete elimination of the possibility of impregnating the specimen with the abrasive, but at the same time retaining a reasonable rate of polishing.

9.2. Polishing abrasives

As with laps there is a considerable choice of abrasives available for polishing. These include

- *silicon carbide (carborundum)* powders in a variety of grade sizes, the finest available being 1200 mesh (3 μm);
- *aluminium oxide* powders from 30 μm to 1 μm in size. Still finer grades are available, but these are not generally used in the preparation of geological specimens, though 0.3 μm alumina is sometimes used for ceramic specimens;

- *boron carbide* powders in grades down to 1000 mesh (5 μm). This material is harder than carborundum, and less expensive than diamond. It can be used for grinding and polishing extremely hard geological materials and ceramics;
- *diamond*, available in grades as small as 0.25 μm and supplied in various forms as suspensions in oil, pastes, and sprays. Despite its high price, it is perhaps the most economical of all the polishing media.

9.3. The choice of lap material and abrasive

The choice of material for a polishing lap depends on a variety of factors, in particular, the type of abrasive, the nature of the lubricating fluid used, as well as the nature of the specimen to be polished.

For ore minerals, or other materials with low porosity, lead laps are commonly used. They are not used for more porous materials since the lead dust produced in polishing is not easily removed from pores or cavities and for these materials paper discs are preferred. High loads are now commonly applied to the specimen in the final stages of polishing and smearing of the grains is best avoided by the use of the non-woven discs made from highly stable synthetic fibres.

The choice of abrasive is essentially restricted to aluminium oxide powders and diamond, though as mentioned above, 1200 grade carborundum will give a moderate polish. For general use, diamond is most commonly employed, since it is less likely to cause smearing of the surface than aluminium oxide.

9.4. A brief outline of the technique of polishing

While hand polishing of specimens is perfectly possible, it is an extremely laborious task, though it can be expedited by the use of simple rotating plates such as were described earlier (Section 3.4) and the chucks used for the initial grinding of rock specimens (see Fig. 6).

The specimen to be polished is cast into a suitable resin (see Sections 8.1 and 8.2) either as a rectangular block or in a short cylindrical mould such as a 10 mm ($\frac{1}{2}$″) length of Perspex tubing 25 mm (1″) in diameter, with an identification label placed in a suitable position. The top and bottom surfaces of the block are ground flat with 240 mesh carborundum and water to form a manageable specimen, with one face of the specimen exposed. Care should be taken to keep the faces parallel. The top surface is then ground with successive grades of carborundum down to 600 or 800 mesh carborundum on a conventional cast-iron rotating lap. Grinding should continue for up to 30 min with the finer grades in order to remove

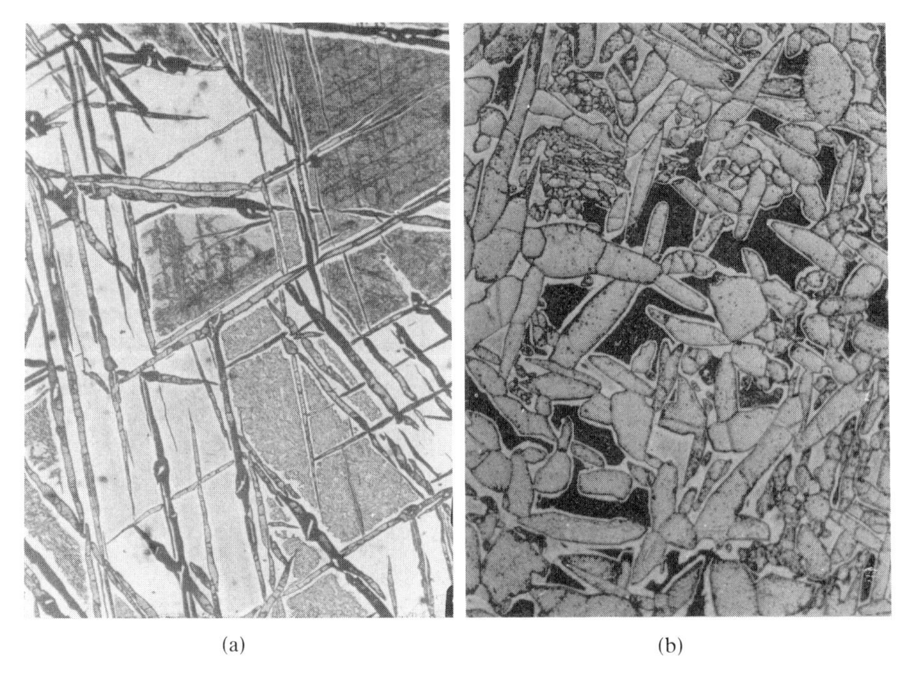

(a) (b)

Fig. 12. Opaque sections of meteorites with cover slips fixed with Canada balsam. Photographed by reflected light using a cover slip as a reflector. The sections were made by H. C. Sorby in 1863. (a) Tazewell meteorite; (b) Pallas meteorite; magnification (a) ×20; (b) ×50.

any micro-structural damage that may have been caused by sawing or earlier stages of grinding. Ideally each stage of grinding should be carried out on a separate machine to minimize the risk of transfer of abrasive from one stage to the next. If this is not possible, the laps should be changed or very thoroughly cleaned between grades, as should the specimen.

Following the carborundum stages, the specimen is transferred to the polishing machine (or the lap changed to the polishing lap) together with a very small quantity of a diamond slurry such as Engis' Hyprez (8 μm) or Logitech diamond paste (6 μm). After about 60 minutes of polishing, the lap is changed and polishing continued with 3 μm, 1 μm, and $\frac{1}{4}$ μm diamond paste, each step continuing for at least 60 minutes* and an increase in speed of rotation of the lap for the last two stages. The lap should be kept moist, but not flooded, with the alcohol-based polishing fluid supplied by the manufacturers of the diamond paste. The specimen is cleaned between each stage, preferably in an ultrasonic bath. The laps should be carefully covered when not in use, since a single initial charge of diamond will polish several specimens with the addition of only minimal amounts of paste unless the specimens are particularly hard.

*Some manufacturers claim that their products will polish in much shorter times. It is up to the user to determine the times which give the most satisfactory results.

The finished specimens should be kept in a sealed container, together with anhydrous silica gel, since the polished surface is not normally protected from the atmosphere by a cover glass.

For low-power examination of polished surfaces by reflected light, the standard objectives of the polarizing microscope will probably prove satisfactory, but at high magnifications, it is, of course, desirable to use non-coverslip corrected objectives as well as a properly constructed reflector unit for illumination of the specimen.

Although polishing is the best method of preparing a specimen for examination by reflected light, it is possible to produce useful slides by simple hand grinding (with 3F carborundum powder) and covering the specimen with a cover slip (Fig. 12). The cement fixing the slip, in effect, fills in the grooves left by the abrasive and at relatively low magnification, these are invisible. With oblique incident light, some unwanted reflections may occur, but this difficulty can readily be overcome by reflecting the beam of a powerful lamp on to the slide with a piece of a clean cover slip mounted above the slide at 45° to it.

10 Polished thin sections and ultra-thin sections

With the development of polishing techniques using diamond pastes, there has been increasing demand for thin sections with either one or both surfaces ground and polished with a minimum of surface relief and for ultrathin sections (10 μm in thickness). Polished thin sections are required especially for electron microprobe and reflective (backscattered) scanning electron microscope examination, while ultrathin sections permit the study of fine-grained rocks with a particle size appreciably less than 30 μm, such as clays, shales, and some limestones, especially chalk, which appear almost opaque in standard sections. Uncovered polished thin sections give good resolution and need neither oil nor glycerine for examination at high magnification, in contrast to sections finished only with 600 mesh carborundum. The practice of not covering slides for research purposes allows other techniques, for example, staining (see Chapter 11) to be easily applied. For teaching purposes covering is essential to protect the section and it is doubtful if polishing is then justified.

The techniques employed are essentially those used to make standard sections combined with the methods of polishing outlined in Chapter 9. While polished and ultra-thin sections can be made by hand, it is more usual to employ machine methods.

10.1. Polished standard thin sections

Hand polishing of one or both surfaces of a standard thin section presents few problems other than the shear tedium of the task.

The initial slice of rock or ceramic must be cut rather thicker for a polished slide than for a standard slide, nearer 8 mm. This is to allow the removal of the glass mounting plate of the first stage of preparation. The slice is smoothed on a fine carborundum plate if necessary and mounted with epoxy resin on a small glass plate made by cutting a standard microscope slide in half. The mounted slice is then ground with 300 mesh carborundum and water or other lubricant if the material is water sensitive, for about 30 minutes, followed by 600 mesh carborundum. If the slide is to be polished on both surfaces, this face is polished using 8 μm diamond paste. Polishing with finer grades on this face appears to have little effect on the improvement in resolution and may, in fact, lead to poor adhesion between the specimen and the resin and lifting of the slice from the glass in later stages of sectioning. Whether polished or not, this face of the specimen is now cemented to the final slide (usually 76 × 26 mm (3″ × 1″) or smaller if required by the subsequent technique of examination) with epoxy resin. The original glass plate is now removed by slicing the specimen parallel to its surface, or the glass can be removed by grinding. The slice is ground down with 300 mesh carborundum to about 300–400 μm and then ground to about 35 μm with 600 mesh carborundum. The specimens and holder are then thoroughly cleaned and transferred to the polishing lap, preferably on a separate machine. The surface is polished using successively finer grades of diamond paste as described in Chapter 9.

As indicated earlier, polished thin sections are not normally covered.

10.2. Ultra-thin sections

Standard thin sections of rocks and ceramics with grain sizes less than about 30 μm are generally very disappointing objects for microscope study because the grains tend to be piled one upon the other within the thickness of the section. The result is a confused mass of particles, none of which can be readily resolved. In fact, for many years the microscopical study of clays, shales, and some ceramics was restricted to the study of the coarser components. The development of polishing methods and the preparation of ultra-thin sections (<10 μm) has gone some way to overcoming this difficulty.

The technique of producing ultra-thin sections requires simply the extension of the time of polishing of the slice with 1 μm diamond paste. This is a very slow process which cannot be hastened and the operator must exercise continuous vigilance if the specimen is not to be lost completely, especially if the rock is soft.

The thickness of the ultra-thin section cannot be measured by the normal method of observing the polarization colours of common minerals; at less than about 15 μm most minerals show black or grey polarization colours. Micrometer measurements must be made frequently or else great reliance placed on the dial gauge attached to the specimen holder. As always in slide making, the slice should be slid sideways off the lap and never lifted vertically, to avoid pulling the very fragile thin section away from the glass mount.

Even thinner sections, 1 or 2 μm thick, can be prepared by ion-beam thinning (Goodhew 1984), a method which involves the bombardment of the specimen with an ion beam, *in vacuo*, which literally burns away successive layers of the section.

The development of the polished thin section, the ultra-thin section, and ion-beam thinning has made possible the qualitative and quantitative examination of rocks and ceramics with the scanning electron microscope, the scanning transmission electron microscope, the electron microprobe, and the cathodoluminescence microscope. Once again, these techniques are beyond the scope of this book and the interested reader is referred to the works of Reed (1975), Long (1977), Goldstein *et al.* (1981), Heinrich (1981), and Marfunin (1979). A brief account of these techniques is given by Fairchild and by Miller in Tucker (1988).

11 Staining, etching, and peels

While many of the minerals in rocks and ceramics can be readily identified in thin sections by optical methods alone, there are several common minerals which are difficult to differentiate from each other. These include, for example, the carbonate minerals in limestones, gypsum and anhydrite, some feldspars in rocks and in ceramics, and some of the components of cement clinkers.

The theory of staining of rocks is not fully understood, but it is clear that the coloured substances formed are only surface coatings on the mineral particles and do not penetrate into the body of the grain. They can readily be removed by abrasion and all operations connected with staining should be carried out gently.

11.1. Carbonate minerals

The simplest method of distinguishing these various minerals has been found to be by the application of stains and by etching with dilute acids. The methods can be applied to hand specimens, and polished surfaces as well as to thin sections. In addition, stained peels can be made quite simply which avoid the necessity for making a thin section, if so desired.

Although there is a very large literature in this field, only a few of the well tried methods are outlined here (Friedman 1959; Davies and Till 1968).

The limestone surface is ground flat and polished with 400 and 800 mesh carborundum and washed thoroughly clean. Alternatively, an uncovered thin section may be used. *This surface must not be touched with the fingers or contaminated with grease in any way.* It is then etched lightly with 1.5% hydrochloric acid. The time required for etching is a matter for experiment: on some limestones no etching is required, on others 60 seconds is adequate. Prolonged etching of a thin section may remove all the calcite. In general, fine grained calcilutites require longer etching than coarse grained calcarenites.

11.1.1. Alizarin red-S and potassium ferricyanide

Two solutions are made up as follows:

alizarin red-S: 0.2 g in 100 ml of 1.5% HCl.

potassium ferricyanide: 2 g in 100 ml of 1.5% HCl.

These two solutions are mixed in the ratio of alizarin red-S:potassium ferricyanide, 3:2, immediately before use. The specimen or thin section is immersed for 40–60 seconds and then washed very gently with distilled water—the stain is merely a surface coating and is very delicate. The intensity of colour can be increased by immersing in the dye solution for 10 seconds following the initial staining.

The staining procedure imparts the following colours to the carbonate minerals: calcite and aragonite—pink to red; ferroan calcite—mauve to royal blue, depending on the percentage of iron present; dolomite remains colourless; ferroan dolomite—pale to deep turquoise.

It will be noted that aragonite and calcite are not differentiated by this stain, but if they are present, as in some Recent carbonates they can be distinguished by the use of Feigl's solution.

11.1.2. Feigl's solution

This stain is made by adding 1 g of solid silver sulphate to a solution of 11.8 g manganese sulphate in 100 ml of water, and boiling. After cooling, the solution is filtered and one or two drops of sodium hydroxide solution added. The precipitate is filtered off after about 2 hours. The solution remains stable if stored in a dark bottle.

The rock surface is etched as before in dilute hydrochloric acid, washed, and immersed in the solution for 2–4 minutes. Finally the rock is gently washed and allowed to dry in the air.

Calcite and dolomite remain colourless, while aragonite is blackened.

11.1.3. Alizarin cyanine green

This stain is used specifically for high magnesian carbonates and dolomite. It is made up of the following two parts:

(1) 0.2 g of the dyestuff dissolved in 25 ml of methanol;
(2) 30 g of sodium hydroxide pellets dissolved in 70 ml of distilled water.

Parts (1) and (2) are mixed in the ratio 5:3 respectively, immediately before use, and the mixture is brought to the boil. The section is immersed in the boiling solution for 5–10 minutes, washed gently in distilled water, and allowed to dry in the air.

The following colours are imparted: calcite and aragonite—very pale green; high magnesian calcite—dark green; dolomite—very dark green.

Attempts to use these stains successively or in combination to give multi-coloured sections generally prove unsuccessful. Stained sections can, however, be photographed, cleaned by careful scrubbing or repolishing, restained, and photographed again. With care in cleaning off the first stain, the two photographs will show essentially the same rock section.

11.2. Gypsum and anhydrite

Friedman (1959) recommends the use of a solution containing 0.1–0.2 g of alizarin red-S in 25 ml of methanol to which is added 50 ml of 5% sodium hydroxide (in water). The specimen is immersed in the cold solution for a few minutes. Gypsum is stained purple, while anhydrite and calcite are unstained. Dolomite is stained a very faint tint of purple.

No stain appears to be specific for anhydrite and it is best identified by a process of elimination.

11.3. Feldspars

CAUTION: All the methods devised for staining feldspars involve the use of hydrofluoric acid. This is an extremely dangerous substance which causes very severe burns, which may not be apparent until some time after contact with the skin. Inhalation of the vapour can be fatal. Gloves should be worn and the use of a fume cupboard is essential. Waste acid must be neutralized before disposal in the drainage system.

The identification of feldspars in igneous and metamorphic rocks is generally based on the presence of twinning, but in sediments this characteristic is commonly absent and it is necessary to distinguish the feldspars (from quartz) by the presence of alteration products or cloudiness in the grains. This is not entirely satisfactory and techniques for staining both alkali and plagioclase feldspars have existed for many years. Unfortunately, these have not always been reliable, the staining being pale or indiscernible, even with saturated solutions and long staining times. The following method devised by Houghton (1980) is claimed to overcome these difficulties.

Three solutions are required:

(1) 0.02 g K-rhodizonate (dipotassium salt of rhodizonic acid) in 30 ml distilled water. This solution should be filtered before use;

(2) saturated solution of sodium cobaltinitrite in water (about 50 g per 100 ml distilled water);

(3) 2–3% solution of barium chloride in distilled water.

The K-rhodizonate solution is unstable and a fresh solution should be made before each staining session; the cobaltinitrite will last for at least six months in a dark bottle and the barium chloride is very stable.

The slab or thin section (about 40 μm thick) is polished on wet 600 mesh silicon carbide paper and then thoroughly cleaned with detergent to remove any surface residue. The surface should not be touched with the fingers. Since the acid will attack glass, slides can be coated with melted wax (candle grease) with a brush, if desired. It is then etched with hydrofluoric acid vapour contained in a small, shallow polythene or Perspex container with supports for the specimen and a close-fitting lid. The acid used should be strong (52–55% HF) and should be replaced after about 45 minutes use. The etching time is about 25–30 seconds.

The specimen is removed from the acid container with flat plastic tongs and immediately placed in the sodium cobaltinitrite solution for about 45 seconds. It is then washed briefly in tap water (in a beaker), allowed to drain, and the excess water blotted from the end of the slide. The specimen is next immersed briefly (about 2 seconds) in the barium chloride solution and washed as before. Finally, several drops of the K-rhodizonate solution are dropped on to the wet specimen and spread uniformly by gently tilting the slide. After a few seconds, or when the plagioclases turn pink, the slide is washed in a beaker of water. If the plagioclase is still grey or pale pink the intensity of colour can be increased by wetting the slide, re-applying the K-rhodizonate stain, and washing. Finally the slide is dried with compressed air.

By this method, alkali feldspars are stained yellow; pure Na-albite remains colourless; other plagioclases stain pale to deep pink in proportion to the amount of calcium in the molecule. Quartz remains unstained. With very fine-grained rocks, there is a tendency for the rhodizonate stain to spread to adjacent grains, making precise identification difficult.

This stain is especially useful for the identification of feldspars in unconsolidated sands. Such grains must, however, be attached to the slide with a very thin layer of adhesive (e.g. Araldite diluted with toluene or photocement diluted with acetone).

11.4. Etching

As noted earlier, etching with dilute hydrochloric acid is a useful means of differentiating between calcite and dolomite in polished thin sections and slabs. The calcite is partially removed by the acid, whereas the dolomite is virtually unaffected and, particularly under the stereoscopic microscope, the latter stands out in high relief.

The technique of etching is relatively simple. The polished surface of a slab or thin section is immersed face downwards (without touching the bottom of the container) in a dilute solution of 1–5% hydrochloric acid for a few seconds and then washed very carefully. Other acids are sometimes used, including acetic acid and disodium-EDTA. The action of these is more gentle and they are less likely to disrupt delicate structures in the limestone. If etching is prolonged the whole of the specimen may disappear. Where particles insoluble in acid are present, it is sometimes useful to remove all the carbonate in order to examine the non-carbonate particles more closely.

Etching is also used in the identification of phases in Portland cement clinker. Here the etchant is a 1% solution of nitric acid in alcohol (ethanol), which acts more like a stain than a true etching agent, as colour changes occur in the components of the clinker depending on the time of contact. For further information the reader is referred to the work of Dorn and Adams (1983).

Apart from the applications noted here, etching plays little part in the identification of the components of rocks and ceramics. It is, however, important in the making of replicas, more commonly known as peels, of rock surfaces.

11.5. Peels

A 'peel' (or replica) is simply a thin coating of a transparent substance which can be applied to the surface of a rock and subsequently 'peeled' off; it then retains some feature of the surface. This may be an impression of the topography (texture) of the etched surface of a rock or it may lift particles of the rock itself or of a stain applied to it.

Peels of etched rock surfaces have long been used to facilitate the study of textures in carbonate rocks. Replicas of textures in other rock types can also be made, though this is rarely done, since it involves the use of hydrofluoric acid. Peels have also been used to lift the particles of stain from a stained surface (a stained peel) to allow easier examination of the distribution of the component grains. A combination of these methods yields a stained peel which has many advantages in the examination of carbonates (and other rocks).

Since it is unnecessary to make a thin section, a peel is much quicker to prepare. Furthermore, it shows greater detail; grain boundaries and matrix fabrics are more distinct and very small grains are readily resolved. Photomicrographs of peels are much sharper than those of thin sections.

Essentially, the making of a peel consists of casting a thin sheet of clear plastic on to a prepared surface of a rock, removing and mounting the plastic, and examining it under the microscope. In practice, the technique requires attention to detail, but is relatively simple.

Two methods of making the replica of a surface have been described. The first method (described by Davies and Till (1968)) used a 7% solution of ethyl cellulose in trichloroethylene which was poured on to a glass plate edged with Sellotape. The solvent evaporated in a few minutes to leave a very thin sheet (30–100 μm) which could be peeled from the glass and then transferred to the rock specimen. While this gave very satisfactory peels, it has been superseded by a method using commercially made cellulose acetate sheet, which is available in various thicknesses from 0.1 mm upwards. While thicker sheets are easier to handle, the 0.1 mm sheet probably gives the best replication. Acetate sheets are soluble in a number of solvents including tetrachloroethane, methyl acetate, diacetone alcohol, ethyl acetate, and acetone. These solvents should not be inhaled and gloves should be worn to prevent contact with the skin.

The rock specimen is first ground and polished to a smooth surface with 400 and 800 grade carborundum and etched with dilute acid, as described earlier. It is then set *almost* horizontally on a bed of Plasticine or modelling clay.

The surface of the rock is flooded with solvent (usually acetone) and a piece of acetate film about 1 cm larger than the specimen is gently lowered on to the solvent. The film should initially float on the liquid, but as the latter evaporates, the film settles on to the rock. If air bubbles occur or the film wrinkles, no attempt should be made to remove it until the solvent has evaporated (30 minutes to 1 hour). The film can then be stripped off, the specimen repolished, and the procedure can be restarted. If no blemishes have occurred the film can be carefully peeled from the specimen, trimmed with scissors and mounted on to a glass slide using narrow strips of adhesive tape (such as that used for repairing manuscripts or sheet music) along the edges. More permanent mounts can be made by covering the peel with a second microscope slide, but the thickness of the slide may preclude the use of high-power objectives with short working distance.

A simple method of making peels using a proprietary rapid-drying cellulose lacquer (such as 'Humbrol' nitrate cellulose dope used for model aircraft) has been described by Abineri (1986, 1989) which overcomes the need to use potentially toxic solvents. The specimen is prepared as in the methods described above. A self-adhesive paper label with a small (10 mm, $\frac{1}{2}$") window cut in it is then fixed to the prepared surface. The lacquer is applied with a small nylon paint brush and allowed to dry for one hour at 35–40 °C under a desk lamp. The dry film can then be removed and mounted on a standard slide under a cover glass held at the edges by adhesive tape.

If particles of the rock adhere to the peel, they should be examined carefully to determine whether they are non-calcareous 'inclusions' in a calcareous rock or merely pieces of the rock pulled out by the peel. If desired, they can be removed by brief immersion in dilute acid.

Peels of stained rocks can be made as indicated above, but here it is essential that the peel lifts every particle of stain. The original stain should not therefore be too intense. Such a section can be restained with a different stain without further treatment. A series of peels can thus be made of the same section which will allow a complete record of the minerals present to be recorded. Stained peels are capable of high resolution and are ideal for modal analysis studies and for studies of rock diagenesis or grain-boundary relationships.

Finally, the slice used for the peels can be mounted and rubbed down to form a thin section to complement the peels.

The preparation of peels of limestones is the most common application of this technique, but it can be applied to any material that can be etched. Even water-sensitive materials could probably be examined in this way, using a rapid wash in water to differentially dissolve the various components. Since the technique of peel making does not require the making of a thin section, peels can be made quickly from polished and etched surfaces and can be examined by transmitted light using a standard microscope without polarizing equipment. The scope for this technique is probably only limited by the imagination of the investigator.

12 The extraction of heavy mineral suites from sands

A word of explanation of the term 'heavy minerals' and a brief note on the methods of extracting them from sands may encourage some microscopists to investigate a now largely neglected aspect of geology.

All rocks consist of 'essential' minerals, that is, minerals by which the rock can be identified, and 'accessory' minerals. The latter are present usually in very small proportion and are rarely seen in thin sections. However, it is possible to concentrate them by virtue of the fact that they have much higher densities than most of the essential minerals. In liquids of density intermediate between the densities of the essential (or light) minerals and those of the accessory (or heavy minerals), the essential minerals will float, while the accessory minerals will sink. Alternative methods of concentration can be used which depend on differences in other physical properties.

In order to concentrate the accessory minerals, it is essential that the individual grains are disaggregated. With fresh igneous and metamorphic rocks, this is almost impossible to achieve; with weathered rocks, crushing will sometimes produce an acceptable separation of the grains, but the method is most readily applied to loose sands.

The accessory minerals of sands include tourmaline, rutile, and zircon; feldspars and mica; opaque iron minerals; and many more.

Early oil geologists, working in south-east Asia on unfossiliferous horizons realized that sands could be differentiated by their heavy mineral content and in the first half of this century, the heavy mineral suites of many sands were investigated. The heavy mineral suites not only act as 'finger-prints' for the identification of a sandstone horizon, they also serve to indicate the source (or provenance) of a sand and thus to indicate something of the geological history of a sand body.

The examination of large numbers of heavy mineral residues can be a tedious occupation, but can, nevertheless, be rewarding. Unfortunately, the rise of the cult of the 'black box' has led to the belief that microscope examination of the heavy residues is unnecessary and that all the minerals present can be identified by automated methods. Such methods will not distinguish between idiomorphic zircons and rounded zircons, each with their own story to tell of its history, nor will they distinguish between the enormous variety of tourmalines present in some sands (Humphries 1961). There is no substitute for microscopical observation.

12.1. Treatment of sample

Although accessory minerals can be extracted from igneous and metamorphic rocks, only the treatment of loose sands is considered here.

The first step is to examine the sand with a hand lens. Large pebbles can be picked out, possibly for thin sectioning or 'half-sectioning' (Section 8.9) and the remainder of the sand passed through a 30 or 60 mesh sieve and then through a fine sieve (170 or 200 mesh). The 30 to 200 mesh fraction can be sieved into additional fractions, if desired. Each fraction is then treated to remove iron coatings from the grains. Dilute hydrochloric acid or citric acid can be used, but care must be taken to avoid the destruction of minerals such as apatite. The fractions are then dried.

12.2. Mineral separation

The standard technique is to separate the quartz from the heavy minerals by the use of 'heavy liquids', either bromoform (specific gravity, s.g. = 2.89) or acetylene tetrabromide (s.g. = 2.97).* Both are expensive, but there are no satisfactory substitutes.

The simplest form of separation apparatus consists of two glass funnels, 50–75 mm (2–3″) in diameter, supported one above the other. Both funnels are fitted with a short length of rubber tube and a pinch-clip (or screw-clip). The upper funnel is filled with the heavy liquid, to within about 10 mm ($\frac{1}{2}$″) of the lip, and a folded filter paper is placed in the lower one. A receiver is placed below the lower funnel. About 10 g or less of the dried sand fraction is placed in the upper funnel and gently stirred. The funnel is covered with a clock glass and allowed to stand for 15–20 minutes, when the heavy minerals will settle to the bottom and the light fraction (quartz and some feldspar) will float. The clip on the upper funnel is opened briefly to wash the heavy minerals into the filter paper in the lower funnel, leaving the bulk of the heavy liquid and the light fraction in the upper funnel. The heavy minerals are washed with alcohol or acetone and set aside to dry. The washings should be kept separate from the pure liquid used for separation. The remaining heavy liquid can be filtered through a fresh filter paper in the lower funnel and returned to the storage bottle. The remaining sand can be washed with acetone or alcohol and the washings retained as before.

The heavy mineral residue is mounted as described earlier. Finally, the dissolved heavy liquid is recovered from the washings by mixing with several volumes of water and separating the heavy liquid (which is insoluble in water) in a separating funnel.

*These liquids and their vapours are toxic. Operations using them should always be carried out in a fume cupboard.

Other liquids, such as methylene iodide (s.g. = 3.32) and Clerici's solution (s.g. = 4.25) have, in the past, been recommended for the separation of minerals of high specific gravity (e.g. Krumbein and Pettijohn 1938). Methylene iodide is, however, very expensive and is nowadays used only for refractive index measurements, while Clerici's solution (a mixture of thallium formate and thallium malonate) is readily absorbed through the skin and is, like all thallium salts, ***extremely poisonous and in no circumstances should it be used for heavy mineral separation.***

12.3. Alternative methods of mineral separation

The simplest alternative to the heavy liquid separation described above is 'panning'. This method is less efficient, but it is certainly much less expensive, the only apparatus required being a small evaporating dish.

About 5 g of clean sand is placed in the dish and, with a little water, is washed to the lip of the dish which is held nearly vertical under a dripping tap. The drops of water are allowed to fall close to the heap of sand, but not directly on it, while at the same time the dish is given a gentle circular motion. This will cause the lighter quartz particles to be washed away, leaving behind the heavy residue. This can be washed into another receptacle and the process repeated until sufficient residue has been accumulated. The best results are obtained with fairly uniform-grained sands or with sieved fractions. To avoid blocking the drain with sand, a bowl should be placed in the sink below the tap in which the sand is trapped, while the water overflows and goes to waste.

The heavy fraction is likely to be contaminated by quartz grains, but these are easily ignored when the dried concentrate is mounted and examined under the microscope.

Jigging and froth flotation have long been used in the mineral processing industry for the concentration of ore minerals, but they have rarely been applied to the separation of heavy mineral suites on a laboratory scale, though there is probably considerable potential for the keen experimenter.

Magnetic methods of separation have proved successful in extracting heavy minerals from sands, but require powerful and specialized apparatus. They cannot be dealt with here as they are beyond the scope of the present text.

12.4. The identification of heavy minerals

The identification of heavy minerals follows much the same lines as the identification of minerals in thin section, but it has to be borne in mind that detrital grains are not uniform in thickness; rarely do they show cleavage traces; twinning is generally absent, and polarization colours are often masked by the colour of the grains. On the other hand, the shape (Fig. 13) and colour of a grain can be important diagnostic features.

Unfortunately, the specialized texts on heavy mineral identification now appear to be out of print, and the interested microscopist must refer to the standard textbooks of mineralogy and crystallography or seek out second-hand copies of such classics as Milner's *Sedimentary Petrography* (1940).

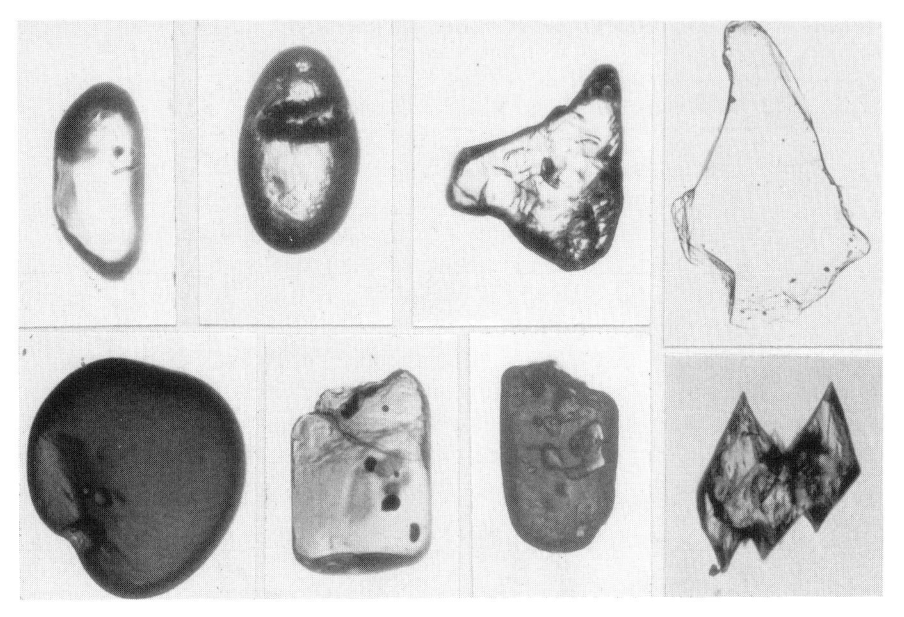

Fig. 13. Heavy mineral grains. Top row, left to right: slightly rounded idiomorphic zircon; well rounded zircon; garnet; mica flake. Bottom row, left to right: well rounded dark tourmaline; slightly rounded idiomorphic tourmaline, fractured after rounding; rounded prismatic tourmaline, fractured after rounding with gas inclusions; staurolite with saw-tooth margins. (All photographs in ordinary unpolarized light; magnification ×200.)

Finale

Though this was intended to be a step-by-step practical guide to the preparation of thin sections of rocks, minerals, and ceramics for the beginner and especially the amateur microscopist, the needs of the young research worker and the undergraduate student user of petrological thin sections has been borne in mind. Inevitably, mention has been made of some of the 'black boxes' that are creeping, or indeed, galloping, into the fields of professional microscopy. But this should not lead to the belief that the basic methods of slide making, initiated by Henry Clifton Sorby 150 years ago, are no longer relevant to the modern world. Whatever electronic device is developed to analyse rocks, minerals, and ceramics, the first requirement is a thin section and the second is an eye (attached to a human brain!) applied to the eyepiece of a polarizing microscope. All the electronic technology in the world will not replace the truly observant eye.

It is doubtful if the reader will have reached this page if he or she is not aware of the purpose of a thin section. On the other hand, he or she might have looked (in the best tradition of the reader of detective stories) at the last page in the hope of finding, not 'who done it', but why it was done.

Rocks, minerals, and ceramics are, by their very nature, solid objects which will yield little to the unaided eye. It is true that chemical analysis will say something about their composition, but this will not tell the whole story of their grain size, their mineral composition, the relationship of one grain to another, the distribution of particular components, or any evidence of the way in which the rock was formed or the ceramic was manufactured. Only when these features have been studied under the petrological microscope will it, perhaps, be time to think of electron microscopes and electron probes. Whether the rocks come from the Moon, from deep in the Earth's crust, or merely from a local quarry, microscopical study of the thin section will provide information unobtainable by other means.

What then of the future of thin-section preparation? As long as we are eager to know more of the origin and history of our world and, one day, other worlds, there will be a place for the slide maker. Even though the hand may have been replaced by machines, the basic methods outlined in these pages will not have changed.

Appendix 1: The petrological microscope and a simple method of adapting a biological microscope for petrological use

The petrological microscope is, in its simplest form, a standard compound microscope with the addition of two polarizing filters, one below and one above the object, a circular rotating stage, and cross-hairs in the eyepiece. An instrument constructed specifically for petrological investigation will have, in addition, strain-free lenses, slots for the insertion of accessory plates, some means of centring the sub-stage, the stage, and the objective about a common axis, and a Bertrand lens for the examination of the rear focal plane of the objective lens. The lower polarizing filter (the polarizer) can be rotated about the optical axis of the microscope and can be swung aside when not required. It also has 'click' stops at 90° intervals which allow the polarizer to be readily set with its plane of vibration in a 'north–south' or 'east–west' direction. The upper polarizing filter (the analyser) is either fitted in a slide in the body of the microscope or arranged as an eyepiece cap. The specimen can thus be viewed in ordinary light, in polarized light, or between crossed polars. The rotating stage is calibrated in degrees and may be rotated through a known angle, the angle being read by means of a fixed point or vernier attached to the microscope stand. The eyepiece is of the focusable type so that cross-hairs (orientated north–south and east–west) can be brought into focus concurrently with the image of the object on the stage.

Although the petrological microscope may appear to be a complex instrument, the principal requirements, namely the polarizing filters (or polars) and the rotating stage, can be attached to most microscopes with a little ingenuity, and a minimum of expense. All that is required is a piece of polarizing film (polaroid), or two photographic polarizing filters (obtainable from the larger photographic equipment dealers), or even a pair of polarizing sunglasses, some stiff cardboard, and a roll of adhesive tape.

A disc of the polarizing film is cut to fit the filter holder of the sub-stage assembly so that it can be swung aside when not required. In the absence of a filter holder, or if a photographic polarizing filter or one lens of the sunglasses is to be used, it can be attached at any convenient place in the light train, beneath the stage. A second piece of polarizing film is cut to fit over the end of a cardboard sleeve which will just slip over the eyepiece. A small photographic filter or the second lens of the sunglasses could just as well be taped to the cardboard sleeve. A plastic, 35 mm film container can be used in place of a cardboard tube, with a hole cut in the end and a piece of polarizing film taped or glued over the hole.

The rotating stage is made from a disc of stiff card rather larger in diameter than the width of the square stage of the microscope, with a central hole of just the right size to accommodate a short cardboard cylinder (made by rolling up a strip of thin card and fixing with adhesive tape) which will fit snugly into the central hole of the fixed stage. If a mechanical stage is fitted to the microscope, it is preferable to remove this, if possible, before attaching the rotating stage. If it cannot be removed, then it may be necessary to make another tube of cardboard just large enough for the stage tube to slide easily in it and attach this second tube to a rectangle of card which will fit the mechanical stage.

It is also perfectly possible to observe some of the features of objects in polarized light by placing the polarizer on the stage of the microscope with the mounted object on top of it and then placing the analyser directly on top of the object. Since only a low-power objective is required initially, there should be plenty of room for this sandwich.

With the simple set-up described here the phenomenon of pleochroism can be observed (provided the object is pleochroic) by rotating the stage or the object over the polarizer, with the analyser removed. To observe birefringence, it is first necessary that the planes of vibration of the polarizer and the analyser are set perpendicular to each other. All this requires is that the analyser is rotated over the polarizer until no light (or only the minimum light) is transmitted. In this position the polars are said to be 'crossed'.

In a standard petrological microscope the plane of vibration of the polarizer is placed in the east–west direction. The simplest method of orientating the polarizer is to look at the reflection from a polished surface through the polar. Reflected light is partially plane-polarized with the vibration direction parallel to the reflecting surface. On rotating the polar, the reflection will vary in intensity. At the position of minimum intensity therefore the vibration direction of the polar is perpendicular to the line of the reflecting surfaces. This is the basis of polarizing sunglasses which are made with the vibration direction vertical so as to eliminate the (horizontally) plane-polarized light (and therefore glare) from reflecting road or water surfaces. Another more precise method is to make use of the pleochroism of a mineral such as biotite, in a thin section of a biotite granite (obtainable from dealers in petrological specimens (see the list of suppliers, p. 73). Biotite may be recognized in thin section by its perfect cleavage, brown colour, and strong adsorption. The direction of vibration of the polarizer is indicated by the direction of the cleavage when the absorption is greatest, i.e. the colour is darkest. Hence, to set the position of the polarizer, the cleavage is placed in the east–west direction and the polarizer is rotated until the biotite shows the maximum absorption. If the thin section is now removed, the analyser can be inserted or placed over the eyepiece and rotated until the field of view is at its darkest. The polars are now in the crossed position. In the simple conversion described above, it is not always possible to fit the eyepiece with cross-hairs and the cardinal points will have to be guessed at.

There are, unfortunately, no simple household items that show pleochroism, but there are several substances wherein isotropism, anisotropism, and birefringence may be readily observed. All that is required is a few grains of common salt, granulated sugar, and sand mounted on a glass slide using a non-aqueous mountant (e.g. glycerine, olive oil, liquid paraffin, or thin mineral oil) and covered with a cover slip. In ordinary light, that is with both polars removed, the common salt will appear as cubes, the sugar will clearly have crystal form but will not be cubes (cane sugar is monoclinic), while the sand grains (quartz) will generally be rounded or perhaps angular. All the grains are clearly visible, despite the fact that they are colourless in a colourless fluid. This is because they all have indices of refraction higher than that of the mounting liquid. If the grains are mounted in clove oil, Canada balsam, or a photocement, which have a refractive index similar to that of the grains, they become almost invisible. This difference in 'relief' is an important characteristic in identifying mineral grains. Since none of these materials are pleochroic, their appearance will not change if the polarizer is put in place and the stage is rotated, but if the analyser is also put in place (in the crossed position) the salt grains will apparently disappear, while the sugar and the sand will stand out bright against the dark background. If the stage is now rotated and careful note taken of individual grains, it will be seen that the grain goes dark or *extinguishes* at intervals of 90°, and between these positions will be illuminated, being brightest at 45° from the position of extinction. The salt grains remain dark as the stage is rotated and are said to be isotropic, whereas the sugar and sand are anisotropic. The colours they show are their *polarization colours* which depend on the *birefringence* of the material and the thickness of the grain.

If the grains of salt and sugar are crushed by rolling them between two glass slides, only their crystal outlines are destroyed—their optical properties remain unchanged. The polarization colours of the sugar will change from white (high-order white) to yellow, blue, and grey (of first or low order) only because the thickness of the grains has been reduced, not because their birefringence has been affected. It is more difficult to crush the sand grains, but they too will show no change in optical properties.

This simple experiment illustrates a few of the features by which minerals can be identified, either as grains, which have been used here for convenience, or in thin sections.

For further details, the reader is referred to the books listed on p. 79.

Appendix 2: Derivation of the equation for the determination of the refractive indices of mounting media (p. 25)

The refractive index of an anisotropic mineral depends on the direction in which it is measured and varies continuously between the minimum and maximum values. In a uniaxial mineral the directions of minimum and maximum refractive index are mutually perpendicular and can be represented as the semi-axes of an ellipse. Intermediate values can therefore be represented by any point on the ellipse and can be determined by calculating the distance of this point from the origin.

The equation of the ellipse is given by

$$\frac{x^2}{a^2} + \frac{y^2}{b^2} = 1$$

where x and y are the coordinates of any point on the ellipse and a and b are the values of the minor and major semi-axes respectively.

If the minor and major semi-axes are represented by the minimum (ε) and maximum (ω) refractive indices, respectively, and if θ is the angle between the positions at which the refractive index of the grain and that of the mountant are the same, then the values of x and y are given by

$$x = N\sin(\theta/2) \qquad y = N\cos(\theta/2)$$

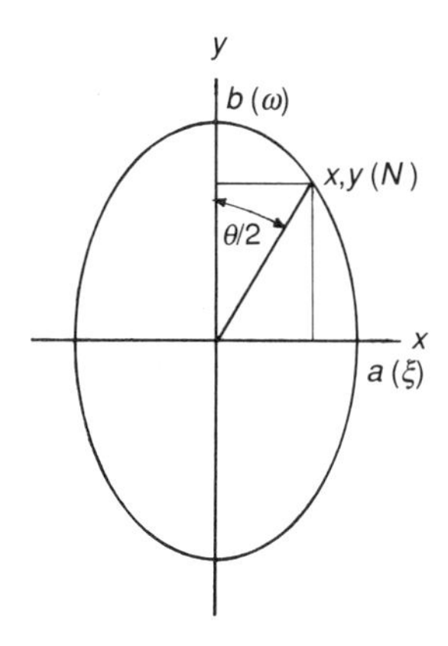

where n is the distance of the point (x, y) from the origin. Substitution in the equation for the ellipse gives

$$\frac{N^2 \sin^2(\theta/2)}{\varepsilon^2} + \frac{N^2 \cos^2(\theta/2)}{\omega^2} = 1$$

and re-arranging gives the required refractive index (N):

$$\frac{\omega^2 N^2 \sin^2(\theta/2) + \varepsilon^2 N^2 \cos^2(\theta/2)}{\varepsilon^2 \omega^2} = 1$$

therefore

$$N = \left(\frac{\varepsilon^2 \omega^2}{\omega^2 \sin^2(\theta/2) + \varepsilon^2 \cos^2(\theta/2)} \right)^{1/2}.$$

Appendix 3: Suppliers

Supplier	Products
Bison Adhesives Rowberry House Copse Cross Street Ross-on-Wye HR9 5PD UK	Combi-Standard epoxy adhesive, Car Window adhesive (photocement)
Ciba-Geigy Plastics Duxford Cambridge CB2 4QA UK	Araldite, Epo-tek, and Versamid
ELE International Eastman Way Hemel Hempstead Herts HP2 7HB UK	hotplates, microscope slides, cover glasses, slide boxes and cabinets, hammers, rock-cutting machines, abrasive blades, diamond blades
Engis Limited Park Wood Trading Estate Sutton Road Maidstone Kent ME15 9NJ UK	diamond powders, silicon carbide discs, lapping discs, lapping machines
Fahrenheit Laboratory Supplies Northfield Road Rotherham S60 1RR UK	Canada balsam, microscope slides, cover glasses, hotplates
Gregory, Bottley & Lloyd 8–12 Rickett Street London SW6 1RU UK	rocks, minerals, fossils, hammers, Lakeside 70C cement, slide cabinets
Hodge, Jacques Abrasives Lymore Avenue Oldfield Park Bath BA2 1AU UK	carborundum powders

Loctite Holdings Ltd Watchmead Welwyn Garden City Herts AL7 1JB UK	superglues, photocements
Logitech Ltd Erskine Ferry Road Old Kilpatrick Glasgow G60 5EU UK	automatic and semi-automatic cutting and grinding machines
Norland Products PO Box 145 North Brunswick NJ 08902 USA	photocement
Palouse Petro Products RT 1, Box 92 Palouse Washington USA	Petropoxy 154
Production Techniques Ltd 13 Kings Road Fleet Hampshire GU13 9AU UK	Petropoxy 154, Lakeside 70 cement, rock-cutting machines
Robnorganic Systems Ltd Highworth Road South Marston Swindon Wilts SN3 4TE UK	mounting media
Washington Mills Electro-Minerals Company (UK) Ltd Mosley Road Trafford Park Manchester M17 1NR UK	carborundum and aluminium oxide abrasive powders

R. G. Widdowson 32 Oak Road Scarborough YO12 4AR UK	fossils, minerals, and rocks, cutting and grinding machines

Note: It is advisable to check prices before ordering. 'Minimum order' charges are applied by some suppliers.

Glossary

Note: Words in *italic* type are defined elsewhere in this glossary.

analyser The upper *polar* in a polarizing light microscope.

anisotropic, anisotropism lit. exhibiting unequal physical properties in different directions. In the majority of transparent minerals the optical properties are dependent on the direction of propagation of light. This leads to many minerals showing bright colours between *crossed polars* (*polarization colours*) or changing colour (or intensity of colour) when rotated in *plane-polarized light* (*pleochroism*). See also *isotropic, isotropism.*

Becke line A bright line surrounding or enclosed by a grain in a mounting medium when the grain is slightly out of focus. It is best seen when the concave side of the microscope mirror is used and the sub-stage diaphragm is partially closed. It is due to total internal refraction of light at the boundary between two substances of different refractive index. As the distance between the objective and the object is increased the ring of light moves from the area of lower refractive index to that of higher refractive index. In thin sections it is best seen in grains at the edge of the slide. In some cases it is possible by this method to compare the refractive indices of two minerals in contact, but the contact is not always sufficiently good to give positive results. It is commonly used to distinguish mineral grains having a refractive index higher than that of the mounting medium, from those with a refractive index lower than that of the mountant. If the grain and the mountant have the same refractive index (e.g. quartz in Canada balsam), no Becke line can be seen. This Becke line test can be extremely useful in discriminating between grains with few other distinguishing features, such as quartz and detrital feldspar (which rarely shows twinning).

biaxial A mineral having two *optic axes.*

birefringence The numerical difference between the maximum and minimum refractive indices of an anisotropic mineral.

chip A fragment of material broken from a larger specimen with a hammer or a fragment produced by a drill (as in well-drilling).

conoscopic observation or observation in *'convergent' light.* The observation of a crystal between *crossed polars* in strongly convergent light (conoscopic illumination), in place of the parallel light normally used in microscopy, allows its optical character to be examined in many directions at the same time. The image observed is variously called the directions image (as opposed to the object image), the image in convergent light, or the *interference figure.*

convergent illumination or **convergent light** A bundle of strongly converging light rays illuminating the object. Strong convergence can be achieved by adjusting the position of the sub-stage condenser or introducing an additional 'bulls-eye' lens above the condenser.

crossed polars The position of the *polarizer* and *analyser* when their planes of vibration are mutually perpendicular. When the medium between the polars is *isotropic* the field of view is dark.

double refraction, doubly refracted In an *anisoptropic* mineral, the division of incident light into two polarized rays with their *planes of vibration* mutually perpendicular.

extinction The condition when a section of an *anisotropic* mineral is dark when rotated between *crossed polars.*

extinction angle The angle between the position of *extinction* of an anisotropic mineral and a cleavage direction or a crystal boundary. The angle can be measured using the calibration of the circular stage of the polarizing microscope.

first-order white, grey, etc. See *Newton's scale of colours.*

flash figure An interference figure formed when the *optic axis* (of a *uniaxial* mineral) or the plane containing both optic axes (*biaxial* mineral) is parallel to the plane of the crystal section. As the stage is rotated, the figure appears momentarily (i.e. it flashes into and out of view) at 90° intervals. In parallel light, these sections show the maximum birefringence.

high-order colour See *Newton's scale of colours.*

interference figure The image formed when an anisotropic crystal, placed between *crossed polars,* is illuminated with *convergent light.* This image cannot ordinarily be observed through the eyepiece because it lies far removed from the focal plane of the latter. It can be observed either by removing the eyepiece and examining the back focal plane of the objective directly through the *analyser* or inserting a small lens (the Bertrand lens) between the eyepiece and the analyser. Only sections of crystals perpendicular to an *optic axis* will give a readily observable interference figure, other sections will yield a *flash figure* or an indistinct or partial figure.

isotropic, isotropism lit. exhibiting equal physical properties in all directions. All cubic minerals and basal sections of tetragonal, trigonal, and hexagonal minerals are isotropic, and are dark when viewed between *crossed* polars. Beware of confusing opaque minerals, which are dark in ordinary light as well as between crossed polars, with isotropic minerals.

Newton's scale of colours A repeating sequence of colours (divided into 'orders') produced when white light passes through a quartz wedge between *crossed polars* and oriented at 45° to their planes of vibration. At the thin end of the wedge the first 'colour' is black followed by grey, white, yellow, orange, red, and violet, then indigo, blue, green, yellow, red. The sequence indigo to red is then repeated. The colours are strongest at the thin end of the wedge and become progressively paler. After three or four bands the colours are very faint and the bands appear almost white (high-order white). The white band at the thin end of the wedge is referred to as first-order white and any colour in this sequence as a low-order colour. These colours depend on the birefringence of quartz and the thickness of the wedge at any given position. Similar colours are produced by sections of colourless *anisotropic* minerals and can be used to determine the *birefringence* of the mineral, provided its thickness is known (and vice versa). A chart (known as a 'Michel-Levy' chart) showing Newton's scale of colours and scales of birefringence and thickness is the usual form in which this relationship is employed in optical mineralogy. It is reproduced in several of the books cited on p. 79.

optic axis The direction of propagation of light within an *anisotropic* mineral along which light is not *doubly refracted*, i.e. the mineral behaves as though it were *isotropic*.

plane-polarized light Light vibrating in only one direction. Often used to refer to the condition for observing a thin section with only the *polarizer* inserted in the light path of the microscope.

plane of vibration The plane containing the vibration direction of a polarized ray and the direction of propagation.

pleochroic, pleochroism lit. more colours. Coloured *anisotropic* minerals which, when viewed in *plane-polarized light*, show changes in colour or brightness in different directions of propagation of light.

polar Any device for producing or analysing *plane-polarized light*.

polarization colours The colours displayed by an anisotropic mineral viewed between *crossed polars*.

polarizer The lower *polar* in a polarized light microscope.

refractive index The ratio of the speed of light in a vacuum to that in a given medium. Normally determined by comparison methods using the *Becke line* test.

slice A parallel sided piece of a specimen produced by sawing.

uniaxial A mineral having one *optic axis*.

Bibliography

Recommended further reading

Adams, A. E., MacKenzie, W. S., and Guilford, C. (1984). *Atlas of sedimentary rocks under the microscope.* Longman, London.

Bradbury, S. and Robinson, P. C. *Qualitative polarized light microscopy*, RMS Microscopy Handbooks No 09. Royal Microscopical Society/Oxford University Press, Oxford. (In press.)

Bradbury, S., Evennett, P. J., Haselmann, H., and Piller, H. (1989). *Dictionary of light microscopy*, RMS Microscopy Handbooks No 15. Royal Microscopical Society/Oxford University Press, Oxford.

Cox, K. G., Price, N. B., and Harte, B. (1974). *An introduction to the practical study of crystals, minerals, and rocks.* McGraw Hill, London.

Cox, K. G., Bell, J. D., and Pankhurst, R. J. (1979). *The interpretation of igneous rocks.* Allen & Unwin, London.

Deer, W. A., Howie, R. A., and Zussman, J. (1974). *An introduction to the rock-forming minerals.* Longman, London.

Gay, P. (1967). *An introduction to crystal optics.* Longman, London.

Hartshorne, N. H. and Stuart, A. (1964). *Practical optical crystallography.* Arnold, London.

Horowitz, H. S. and Potter, P. E. (1971). *Introductory petrography of fossils.* Springer, Berlin.

MacKenzie, W. S. and Guilford, C. (1981). *Atlas of rock-forming minerals in thin section.* Longman, London.

MacKenzie, W. S., Donaldson, C. H., and Guilford, C. (1982). *Atlas of igneous rocks and their textures.* Longman, London.

Smith, H. G. (1940). *Minerals and the microscope* (4th edn). Murby, London.

References

Abineri, K. W. (1986). The preparation of cellulose lacquer rock peels. *Microscopy* (*The Journal of the Queckett Microscopical Club*), **35**, 451–9.

Abineri, K. W. (1989). Photomicrographs of cellulose peels from the Mesozoic rocks of Dorset. *Proc. Geol. Assoc.*, **100**, 161–74.

Brewster, D. (1837). *A treatise on the microscope.* A and C Black, Edinburgh.

Davies, P. J. and Till, R. (1968). Stained dry cellulose peels of Ancient and Recent impregnated carbonate sediments. *J. Sedimentary Petrology*, **38**, 234–7.

Delly, J. G. (1988). Inexpensive orange filter for refractive index determination. *Microscope*, **36**, 213–15.

Dorn, J. D. and Adams, L. D. (1983). The etch rate of Portland cement clinkers as it relates to structure and hydraulic potential. *Microscope*, **31**, 37–42.

Fox Talbot, W. H. (1834*a*). On Mr. Nicol's polarizing eyepiece. *Phil. Mag.*, 289–90.

Fox Talbot, W. H. (1834*b*). Experiments on light. ibid., 321–34.

Friedman, G. M. (1959). Identification of carbonate minerals by staining methods. *J. Sedimentary Petrology*, **29**, 87–97.

Goldstein, J. I., Newbury, D. E., Echlin, P., Joy, D. C., Fiori, C., and Lifshin, E. (1981). *Scanning electron microscopy and X-ray microanalysis.* Plenum, New York.

Goodhew, P. (1984). *Specimen preparation for transmission electron microscopy of materials*, RMS Microscopy Handbooks No 03. Royal Microscopical Society/ Oxford University Press, Oxford.

Heinrich, K. F. J. (1981). *Electron beam X-ray microanalysis.* Van Nostrand Reinhold, New York.

Houghton, H. F. (1980). Refined techniques for staining plagioclase and alkali feldspar in thin section. *J. Sedimentary Petrology*, **50**, 629–31.

Humphries, D. W. (1961). The Upper Cretaceous White sandstone of Loch Aline, Argyll, Scotland. *Proc. Yorks. Geol. Soc.*, **33**, 47–76.

Jones, T. and Hawes, R. W. M. (1964). The preparation of ultra-thin petrological sections by semi-automatic methods. *The Microscope and Crystal Front*, **14**, 200–8.

Krumbein, W. C. and Pettijohn, F. J. (1938). *Manual of sedimentary petrography.* Appleton-Century-Croft, New York.

Long, J. V. P. (1977). Electron probe microanalysis. In *Physical methods in determinative mineralogy* (2nd edn) (ed. J. Zussman). Academic Press, London.

Marfunin, A. S. (1979). *Spectroscopy, luminescence and radiation centres in minerals.* Springer, Berlin.

Milner, H. B. (1940). *Sedimentary petrography.* Murby, London.

Porguen, V. (1989). A high index fixative for mounting microfossils. *Microscope*, **37**, 58–62.

Reed, S. J. B. (1975). *Electron microprobe analysis.* Cambridge University Press.

Sorby, H. C. (1851). On the microscopical structure of the Calcareous Grit of the Yorkshire coast. *Quart. J. Geol. Soc. Lond.*, **VII**, 1–6.

Sorby, H. C. (1868). In Beale, L. S. *How to work with the microscope* (4th edn), pp. 179–83. Harrison, London.

Sorby, H. C. (1882). Preparation of transparent sections of rocks and minerals. *The Northern Microscopist*, **2**, 101–40.

Tucker, M. (ed.) (1988). *Techniques in sedimentology.* Blackwell, Oxford.

Weatherhead, A. V. (1947). *Petrographic micro-technique.* Arthur Barron, London.

Index